AF576023

Crystallographic Studies of Enzymes (Volume II)

Crystallographic Studies of Enzymes (Volume II)

Editors

Kyeong Kyu Kim
T. Doohun Kim

MDPI • Basel • Beijing • Wuhan • Barcelona • Belgrade • Manchester • Tokyo • Cluj • Tianjin

Editors
Kyeong Kyu Kim
Precision Medicine
Sungkyunkwan University
School of Medicine
Suwon
Korea, South

T. Doohun Kim
Chemistry
Sookmyung Women's
University
Seoul
Korea, South

Editorial Office
MDPI
St. Alban-Anlage 66
4052 Basel, Switzerland

This is a reprint of articles from the Special Issue published online in the open access journal *Crystals* (ISSN 2073-4352) (available at: www.mdpi.com/journal/crystals/special_issues/Enzymes_II).

For citation purposes, cite each article independently as indicated on the article page online and as indicated below:

LastName, A.A.; LastName, B.B.; LastName, C.C. Article Title. *Journal Name* **Year**, *Volume Number*, Page Range.

ISBN 978-3-0365-6072-4 (Hbk)
ISBN 978-3-0365-6071-7 (PDF)

Cover image courtesy of Kyeong Kyu Kim

Contents

About the Editors

Kyeong Kyu Kim

Kyeong Kyu Kim is a Professor at Sungkyunkwan University School of Medicine. He achieved his B.S., M.S., and Ph.D. degrees at Seoul National University. He worked at UC Berkeley for three years as a post-doc. Then, he moved to Sungkyunkwan University in 2000. His research interests are in the structural and functional understanding of noncanonical DNAs and enzymes, and their chemical modulation for the diagnostic and therapeutic purposes.

T. Doohun Kim

T. Doohun Kim was a Professor at Sookmyung Women's University. He achieved his B.S., M.S., and Ph.D. degrees at Seoul National University. He worked as a post-doc for two years at Seoul National University, and for an additional two years at Harvard Medical School. He started his professor career at Ajou University in 2005, and held the position of Professor until 2015, when he became Assistant and Associate professors. Then, he moved to the Department of Chemistry, Sookmyung Women's University, in 2015, and continued to hold his position of Professor until his death in 2022. His main research interests were the structural and functional understanding of esterases and their industrial applications. In addition, he studied alpha-synuclein and its pathological implications in Parkinson's disease.

Preface to "Crystallographic Studies of Enzymes (Volume II)"

Professor Doohun Kim and I worked as co-editors of the Special Issue of *Crystals*, entitled "Crystallographic studies of enzymes", in 2019, because we believed that it is necessary to collect key achievements in the field of enzyme crystallography for the purpose of sharing the state of the art and thus boosting enzyme research. After the successful collection of 12 articles in the first volume, we continued our effort by editing the second volume. However, we decided to expand the scope of the Special Issue to a functional study rather than limiting it to a crystallographic study. As a result, in the second Special Issue, "Crystallographic Studies of Enzymes (Volume II)", we were able to publish 11 research papers on the structural and functional aspects of enzymes. We strongly believe that the 23 papers collected in these Special Issues will not only provide starting points for those starting to conduct research in this filed, but also give insights to the experts in this field. Unfortunately, Prof. Doohun Kim passed away before finishing this Special Issue. However, I believe that his passion and insights into the structural and functional studies of enzymes makes it possible to finish editing this Special Issue and his contributions will long be remembered.

Kyeong Kyu Kim and T. Doohun Kim

Editors

 MDPI

Editorial

Crystallographic Studies of Enzymes (Volume II)

T. Doohun Kim [1,*] and Kyeong Kyu Kim [2,*]

1 Department of Chemistry, College of Natural Science, Sookmyung Women's University, Seoul 04310, Korea
2 Department of Precision Medicine, Sungkyunkwan University School of Medicine, Suwon 16419, Korea
* Correspondence: doohunkim@sookmyung.ac.kr (T.D.K.); kyeongkyu@skku.edu (K.K.K.)

Citation: Kim, T.D.; Kim, K.K. Crystallographic Studies of Enzymes (Volume II). *Crystals* **2022**, *12*, 1402. https://doi.org/10.3390/cryst12101402

Received: 21 September 2022
Accepted: 22 September 2022
Published: 4 October 2022

Enzymes play a major role in the control of key biological processes by accelerating chemical reactions. It is for this reason that examining their structures and reaction mechanisms is essential for understanding not only the biological processes at a molecule level, but also their application in various fields, such as protein engineering and drug development. After the successful publication of twelve research articles in "Crystallographic Studies of Enzymes", we continue with the series. In the second Special Issue, "Crystallographic Studies of Enzymes (Volume II)", eleven research papers on the structural and functional aspects of enzymes were collected. A brief summary of each article is provided here. While collecting edited articles for this issue, Prof. Doohun Kim, the main editor, passed away. Without his contributions and efforts, it would not have been possible to complete this issue. His passion and insights into the structural and functional studies of enzymes are commemorated in this issue.

Dr. Doohun Kim at Sookmyung Women's University and Drs. Han-Woo Kim, Hackwon Do and Jun Hyuck Lee at the Korea Polar Research Institute published several papers on esterase on a continuum of their previous work [1–3]. They identified new esterases from various microbial sources and characterized their unique enzymatic properties. Furthermore, preliminary crystallographic studies of those esterases were provided [1–3]. In addition, they extended their research scope to the single-stranded DNA that binds protein from the psychrophilic bacterium, *Lacinutrix jangbogonensis*, and this provides an insight into the reaction mechanism of the cold-active enzymes, as well as novel strategies for protein engineering and the application on molecular biological techniques [4].

Drs. Hui-Woong Choe and Young Ju Kim reported the crystallization parameters that affect the space group and diffraction qualities of Ribulose-1,5-Bisphosphate Carboxylase/Oxygenase (RuBisCO) [5]. Based on their observation, a systematic approach required for improving the crystal quality of protein enzymes has been proposed.

Glideosome-associated connector protein (GAC) is a large cytosolic protein (286 kDa) protein that has a role in connecting the parasite F-action with transmembrane adhesin. To provide structural insights on the GAC interaction, Matthews et al. reported the studies on the secondary structure and crystallographic analysis of GAC from *Toxoplasma gondii* [6].

MobB is involved in the biosynthesis of the molybdenum cofactor present in many redox enzymes. Choe et al. proposed that MobB works as an enhancer of MobA activity, based on their structural and biochemical analyses of MobB *Bacillus subtilis* [7].

Schilde et al. reported a systematic analysis of the cross-linked enzyme crystals (CLECs), using halohydrin dehalogenase as an example [8]. They finally concluded that CLECs are suitable for industrial usage since they show good catalytic activities with enhanced mechanical properties.

Nit2, belonging to the nitrilase-like (Nit) branch of the nitrilase superfamily, is known to serve various functions, including as an amidase and a tumor suppressor. In this study, Chang et al. report the crystal structure of Nit2 from *Kluyveromyces lactis* [9]. Based on the structural comparison with other similar structures, they propose a structural relationship among broad spectrum nitrilases.

Immune-responsive gene 1 (IRG1) is an enzyme that plays a role in producing itaconate, a multifunctional immune-metabolite. Dr. Park solved the high-resolution crystal structure of the active-site mutant IRG1 from *Bacillus subtilis* [10]. Based on a structural comparison with the wild-type protein, the author proposes the working mechanism of IRG1.

Wolny et al. in this study performed pico- and nanoscale molecular dynamic (MD) simulations using the high-resolution structure of Hyp-1, a pathogenesis-related class 10 (PR-10) protein from the medicinal herb *Hypericum perforatum*, and analyzed various structural parameters [11]. Based on the study, the authors concluded that MD methods can be used to verify experimental protein models and explain the structural ambiguities.

Author Contributions: Conceptualized, T.D.K. and K.K.K.; writing, K.K.K. Although T.D.K. cannot agree on the publication due to his absence in this world, I believe that all authors including him agreed to the published version of the manuscript. All authors have read and agreed to the published version of the manuscript.

Funding: This research was funded by National Research Foundation of Korea (2021M3A914022936).

Data Availability Statement: Data will be available upon request.

Conflicts of Interest: The authors declare no competing financial interests.

References

1. Hwang, J.; Jeon, S.; Lee, M.J.; Yoo, W.; Chang, J.; Kim, K.K.; Lee, J.H.; Do, H.; Kim, T.D. Identification, Characterization, and Preliminary X-ray Diffraction Analysis of a Novel Esterase (Sc Est) from Staphylococcus chromogenes. *Crystals* **2022**, *12*, 546. [CrossRef]
2. Do, H.; Wang, Y.; Lee, C.W.; Yoo, W.; Jeon, S.; Hwang, J.; Lee, M.J.; Kim, K.K.; Kim, H.W.; Lee, J.H.; et al. Sequence Analysis and Preliminary X-ray Crystallographic Analysis of an Acetylesterase (Lg EstI) from Lactococcus garvieae. *Crystals* **2021**, *12*, 46. [CrossRef]
3. Jeon, S.; Hwang, J.; Yoo, W.; Chang, J.W.; Do, H.; Kim, H.W.; Kim, K.K.; Lee, J.H.; Kim, T.D. Purification and Crystallographic Analysis of a Novel Cold-Active Esterase (*Ha*Est1) from Halocynthiibacter arcticus. *Crystals* **2021**, *11*, 170. [CrossRef]
4. Choi, W.; Son, J.; Park, A.; Jin, H.; Shin, S.C.; Lee, J.H.; Kim, T.D.; Kim, H.W. Identification, Characterization, and Preliminary X-ray Diffraction Analysis of a Single Stranded DNA Binding Protein (LjSSB) from Psychrophilic Lacinutrix jangbogonensis PAMC 27137. *Crystals* **2022**, *12*, 538. [CrossRef]
5. Choe, H.W.; Kim, Y.J. New Insight into the Effects of Various Parameters on the Crystallization of Ribulose-1,5-Bisphosphate Carboxylase/Oxygenase (RuBisCO) from Alcaligenes eutrophus. *Crystals* **2022**, *12*, 196. [CrossRef]
6. Kumar, A.; Zhang, X.; Vadas, O.; Stylianou, F.A.; Pacheco, N.D.S.; Rouse, S.L.; Morgan, M.L.; Soldati-Favre, D.; Matthews, S. Secondary structure and X-ray crystallographic analysis of the Glideosome-Associated Connector (GAC) from Toxoplasma gondii. *Crystals* **2022**, *12*, 110. [CrossRef]
7. Kim, D.; Choi, S.; Kim, H.; Choe, J. Structural and Biochemical Studies of Bacillus subtilis MobB. *Crystals* **2021**, *11*, 1262. [CrossRef]
8. Kubiak, M.; Staar, M.; Kampen, I.; Schallmey, A.; Schilde, C. The Depth-Dependent Mechanical Behavior of Anisotropic Native and Cross-Linked HheG Enzyme Crystals. *Crystals* **2021**, *11*, 718. [CrossRef]
9. Jin, C.; Jin, H.; Jeong, B.C.; Cho, D.H.; Chun, H.S.; Kim, W.K.; Chang, J.H. Crystal Structure of Nitrilase-Like Protein Nit2 from Kluyveromyces lactis. *Crystals* **2021**, *11*, 499. [CrossRef]
10. Park, H. Crystal Structure of an Active Site Mutant Form of IRG1 from Bacillus subtilis. *Crystals* **2021**, *11*, 350. [CrossRef]
11. Smietanska, J.; Kozik, T.; Strzalka, R.; Buganski, I.; Wolny, J. Molecular Dynamics Investigation of Phenolic Oxidative Coupling Protein Hyp-1 Derived from Hypericum perforatum. *Crystals* **2021**, *11*, 43. [CrossRef]

Article

Identification, Characterization, and Preliminary X-ray Diffraction Analysis of a Novel Esterase (*Sc*Est) from *Staphylococcus chromogenes*

Jisub Hwang [1,2,†], Sangeun Jeon [3,†], Min Ju Lee [1], Wanki Yoo [4], Juwon Chang [3], Kyeong Kyu Kim [4], Jun Hyuck Lee [1,2], Hackwon Do [1,2,*] and T. Doohun Kim [3,*]

1 Research Unit of Cryogenic Novel Material, Korea Polar Research Institute, Incheon 21990, Korea; hjsub9696@kopri.re.kr (J.H.); minju0949@kopri.re.kr (M.J.L.); junhyucklee@kopri.re.kr (J.H.L.)
2 Department of Polar Sciences, University of Science and Technology, Incheon 21990, Korea
3 Department of Chemistry, Graduate School of General Studies, Sookmyung Women's University, Seoul 04310, Korea; sangeun94@sookmyung.ac.kr (S.J.); cjw1423@naver.com (J.C.)
4 Department of Molecular Cell Biology, Samsung Biomedical Research Institute, Sungkyunkwan University School of Medicine, Suwon 440746, Korea; vlqkshqk61@skku.edu (W.Y.); kyeongkyu@skku.edu (K.K.K.)
* Correspondence: authors: hackwondo@kopri.re.kr (H.D.); doohunkim@sookmyung.ac.kr (T.D.K.)
† These authors contributed equally to this work.

Abstract: Ester prodrugs can develop novel antibiotics and have potential therapeutic applications against multiple drug-resistant bacteria. The antimicrobial activity of these prodrugs is activated after being cleaved by the esterases produced by the pathogen. Here, novel esterase *Sc*Est originating from *Staphylococcus chromogenes* NCTC10530, which causes dairy cow mastitis, was identified, characterized, and analyzed using X-ray crystallography. The gene encoding *Sc*Est was cloned into the *p*VFT1S vector and overexpressed in *E. coli*. The recombinant *Sc*Est protein was obtained by affinity and size-exclusion purification. *Sc*Est showed substrate preference for the short chain length of acyl derivatives. It was crystallized in an optimized solution composed of 0.25 M ammonium citrate tribasic (pH 7.0) and 20% PEG 3350 at 296 K. A total of 360 X-ray diffraction images were collected at a 1.66 Å resolution. *Sc*Est crystal belongs to the space group of $P2_12_12_1$ with the unit cell parameters of a = 50.23 Å, b = 68.69 Å, c = 71.15 Å, and α = β = γ = 90°. Structure refinement after molecular replacement is under progress. Further biochemical studies will elucidate the hydrolysis mechanism of *Sc*Est. Overall, this study is the first to report the functional characterization of an esterase from *Staphylococcus chromogenes*, which is potentially useful in elaborating its hydrolysis mechanism.

Keywords: carboxylesterase; *Staphylococcus chromogenes*; X-ray crystallography

Citation: Hwang, J.; Jeon, S.; Lee, M.J.; Yoo, W.; Chang, J.; Kim, K.K.; Lee, J.H.; Do, H.; Kim, T.D. Identification, Characterization, and Preliminary X-ray Diffraction Analysis of a Novel Esterase (*Sc*Est) from *Staphylococcus chromogenes*. *Crystals* **2022**, *12*, 546. https://doi.org/10.3390/cryst12040546

Academic Editor: Waldemar Maniukiewicz

Received: 27 February 2022
Accepted: 2 April 2022
Published: 13 April 2022

1. Introduction

Multiple drug resistance (MDR) bacteria are an emerging global threat that potentially imposes healthcare and economic issues [1,2]. The production of drug-inactivating enzymes, such as β-lactamase and aminoglycoside modifying enzymes [3], drug elimination from the cell, mutation of an existing target, and acquisition of a target by-pass system have been proposed as major MDR resistance mechanisms. Therefore, the necessity for discovering and developing novel antibiotics with unconventional modes of action has increased in order to overcome these resistance mechanisms [4].

One of the strategies to avoid MDR is antibacterial prodrugs that are pharmacologically inactive and are cleaved by bacterial enzymes to become active antibiotics [5]. Antibacterial prodrugs are synthesized by adding functional groups to the antibiotic skeleton and may have multiple advantages [5]. For example, adding a lipophilic pivaloyloxymethyl to cephalosporin cefditoren increases its absorption in the small intestine [6]. Ester is also a functional group that is added to antibiotics to increase the delivery efficiency, cell permeability, and oral bioavailability of the prodrug [7,8]. Carbenicillin, carfecillin (phenyl ester), and carindacillin (indanyl ester) are some ester-containing antimicrobial prodrugs [9]. Pathogen specificity is another advantage of ester prodrugs. Since such antibacterial prodrugs are transformed by the cytosolic esterase specifically produced by the pathogen, the pathogen is selectively executed [10].

Previously, human esterases were studied for their function in prodrug activation [11]. However, the application of human esterase for antibiotic prodrug activation is limited due to its esterase-dependent localization and expression. Alternatively, analyzing the substrate selectivity and activity of bacterial esterases has provided crucial details for targeting potential antibiotic prodrugs to develop novel antibiotics for the treatment of MDR [5,7,10,12]. Bacterial esterases have a canonical α/β-hydrolase fold that consists of a core β-sheet surrounded by α-helices to catalyze the hydrolysis (EC 3.1.1.X) of a variety of substrates containing ester groups. The esterases use a catalytic triad comprising a nucleophilic serine, a base histidine, and an activating acidic residue (Asp/Glu) to catalyze the hydrolysis of the ester to a carboxylic acid and alcohol. Despite having the same configuration as the enzyme hydrolase and a high degree of sequence homology, esterases have distinct substrate specificities [13–15]. Therefore, pathogenic esterases need to be functionally investigated, whereas the biochemical and structural studies may provide valuable information for designing species-specific antimicrobial ester prodrugs. This preliminary study focuses on the substrate specificity and function of esterases derived from pathogens. Herein, we have analyzed the distribution of esterases and lipases across the genome of *Staphylococcus chromogenes* NCTC10530, the prevalent bacterial pathogen causing dairy cow mastitis. Furthermore, the carboxylesterase annotated as *Sc*Est has been purified, its biochemical properties have been investigated, and preliminary X-ray studies have been conducted.

2. Materials and Methods

2.1. Phylogenetic Analysis

The subfamily of *Sc*Est was analyzed using a phylogenetic tree based on full-length protein sequences of several lipolytic enzymes that are already classified into specific subfamilies [16–18]. A total of 69 protein sequences, including *Sc*Est and other proteins from the *S. chromogenes strain* NCTC10530 were used for multiple sequence alignment using ClustalX [19]. The neighbor-joining method was used to generate a phylogenetic tree using the MEGA-X [20].

2.2. Gene Cloning, Expression, and Purification of Recombinant ScEst Protein

The gene encoding *Sc*Est (GenBank ID: SUM13810) was amplified by PCR and cloned into the *p*VFT1S plasmid between the *BamHI* and *Xho*I restriction sites. The cloned sequence was verified using Sanger sequencing using T7 promoter and terminator primers. *E. coli* BL21 (λDE3) was transformed with the recombinant plasmid harboring N-terminal 6xHis-tagged *Sc*Est for protein overexpression (Table 1). A single colony from the Luria Bertani (LB) agar plate containing kanamycin was inoculated as a seed culture and grown overnight. The seed culture (20 mL) was inoculated into 1 L of culture medium and kanamycin (50 μg mL^{-1}) and incubated at 37 °C at 150 rpm. When the OD_{600} of the culture reached 0.4, protein overexpression was induced by adding 1.0 mM isopropyl β-D-1-thiogalactopyranoside (IPTG). The cells were further incubated at 37 °C for 4 h, and harvested by centrifugation at 6000$\times$ g. The cell pellets were resuspended in a lysis buffer (20 mM Tris-HCl [pH 8.0], 500 mM NaCl, and 20 mM imidazole) and disrupted by sonication (Vibra-Cell™, Sonics & Materials, Inc., Danbury, CT, USA) for 30 min at 35% amplitude (on for 2 s and off for 4 s). The soluble fraction of protein was separated by centrifugation at 20,000$\times$ g for 40 min.

Recombinant *Sc*Est was purified via a two-step purification process. First, the His-tag-based purification was performed using a His-trap™ FF column (GE Healthcare, Chicago, IL, USA). The supernatant containing the recombinant *Sc*Est was loaded onto the column, and the resin was washed with 10 column volumes of washing buffer. The remaining recombinant *Sc*Est was eluted with two column volume elution buffer (20 mM Tris-HCl [pH 8.0], 500 mM NaCl, 300 mM imidazole). The elute was then concentrated to 5 mL and treated with thrombin for three days at 4 °C in a rotating incubator to cleave the His-tag. For the second purification, HiPrep™ Sephacryl® S-200 HR (Cytiva, Marlborough, MA, USA) connected to an ÄKTA™ Start chromatography system (GE Life Sciences, Piscataway, NJ, USA) was equilibrated with a buffer composed of 20 mM Tris-HCl (pH 8.0), 200 mM NaCl, and 1 mM EDTA, and the protein sample was loaded onto the column. The column was calibrated using cytochrome C (12.4 kDa), carbonic anhydrase (29 kDa), alcohol dehydrogenase (150 kDa), and β-amylase (200 kDa). K_{av} was calculated by $(V_s - V_o)/(V_t - V_o)$, where vs. = elution volume, V_o = column void volume, V_t = column volume. The purity and concentration of the recombinant *Sc*Est were validated using SDS-PAGE and the Bradford protein assay, respectively.

Table 1. Recombinant *Sc*Est protein attributes.

***Sc*Est**	
Source organism	*Staphylococcus chromogenes* strain NCTC10530
DNA source	Genomic DNA
Cloning vector	pVFT1S
Expression host	*Escherichia coli* BL21(DE3)
Amino acid sequence	MKIQLPKPFLFEEGKRAVLLLHGFTGNSSDVRQLG RFLQKKGYTSYAPHYEGHAAPPEEILKSSPHVWY KDALDGYDYLVDKGYDEIAVAGLSLGGVFALKLS LNRDVKGIVTMCSPMYIKTEGSMYEGVLEYARNF KKYEGKDETTIEREMQAFHPTSTLRELQETIQSV RDHVEDVIEPLLVIQAEQDEMINPDSANVIYNEA ASDEKHLSWYKNSGHVITIDKEKEDVFEEVYQFLESLDWSE

2.3. *Enzymatic Analysis*

The substrate specificities of *Sc*Est were measured using various *p*-nitrophenyl esters, including *p*-nitrophenyl acetate (*p*NP-C_2), *p*-nitrophenyl butyrate (*p*NP-C_4), *p*-nitrophenyl hexanoate (*p*NP-C_6), *p*-nitrophenyl octanoate (*p*NP-C_8), and *p*-nitrophenyl decanoate (*p*NP-C_4), obtained from Sigma-Aldrich (St. Louis, MO, USA). The esterase activity with acyl carbon chains of various lengths was evaluated by monitoring the *p*-nitrophenol (*p*NP) in the solution spectrophotometrically [21]. Storage buffer (1 mL) containing 20 mM Tris-HCl (pH 8.0), 200 mM NaCl, and 1 μg *Sc*Est was prepared, and the reaction was initiated by mixing an equal volume of the substrate (final 0.12 μM). The final concentration of acetonitrile in the reaction mixture kept to 5% to avoid micelle formation of substrates with longer acyl chains. The enzyme reactions were analyzed at 405 nm using an Epoch™ 2 microplate spectrophotometer (BioTek Instruments, Winooski, VT, USA), using the storage buffer as control. Three independent measurements were used to represent the activity data.

2.4. *Crystallization, Data Collection, and Structural Analysis*

Commercially available crystallization solutions, MCSG I-IV (Anatrace Inc., Maumee, OH, USA), and JCSG™ and PGA Screen™ (Molecular Dimensions Inc., Altamonte Springs, FL, USA) were used to screen the crystallization conditions of *Sc*Est. The sitting-drop vapor diffusion method was set up by mixing 300 nL of solution and an equal volume of protein (25 mg mL^{-1}) against 80 μL of solution in the reservoir using a mosquito® liquid-handling robot (TTP Labtech Ltd., Hertfordshire, UK). Subsequently, multiple optimizations using 24-well plates were further carried out to obtain a decent size and quality of crystals. The crystallization data are presented in Table 2.

The single crystal of *Sc*Est was cryoprotected using a mixture of crystallization solution where the crystal of ScEst grew and glycerol (25% *w*/*v*) to prevent the crystal from being frozen under a liquid nitrogen stream. The crystal was then mounted on a sample holder. A total of 360 diffraction images were collected at the synchrotron Beamline 7A of the

Pohang Accelerator Laboratory (PAL, Pohang-si, Korea) by rotating at 1° oscillation per frame. The dataset was indexed, integrated, and scaled using the HKL-2000 software package (HKL Research Inc., Charlottesville, VA, USA). The phase of the *Sc*Est structure was successfully determined using the carboxylesterase Est30 (PDB code: 1TQH) with the molecular replacement method. The X-ray diffraction results are listed in Table 3.

Table 2. Initial crystallization conditions and optimization method.

Method	Vapor Diffusion
Plate type for screening	96-well sitting drop MRC plate (Molecular dimension, UK)
Composition of reservoir solution	0.2 M Ammonium citrate tribasic (pH 7.0), 20% PEG 3350
Plate type for optimization	24-well hanging drop plate, (Molecular dimension, UK)
Composition of optimal solution	0.25 M Ammonium citrate tribasic (pH 7.0), 20% PEG 3350
Temperature (K)	296
Protein concentration (mg/mL)	4.3
Composition of protein solution	20 mM Tris-HCl (pH 8.0), 200 mM NaCl
Volume and ratio of drop (protein: solution)	2.0 μL, 1:1
Volume of reservoir (μL)	500

Table 3. X-ray diffraction data.

Data Collection	
Wavelength (Å)	0.9793
X-ray source	PAL 7A
Rotation range per image (°)	1
Exposure Time (s)	1
Space group	$P2_12_12_1$
Unit-cell parameters (Å, °)	a = 50.23, b = 68.69, c = 71.15 α = 90, β = 90, γ = 90
Resolution range (Å) [a]	50–1.66 (1.69–1.66)
No. of observed reflections [a]	402,244 (19,564)
No. of unique reflections [a]	29,406 (1471)
Completeness (%) [a]	99.3 (100)
Redundancy [a]	13.7 (13.3)
R_{sym} [a,b]	0.112 (1.202)
R_{meas} [c]	0.117 (1.250)
I/σ [a]	62.3 (4.0)
CC(1/2) (%)	99.6 (82.8)
Wilson B factor (Å^2)	24.66
Matthews coefficient	2.18

[a] Values in parentheses correspond to the highest-resolution shells. [b,c] $R_{sym} = _{hi} | I(h)_i - \langle I(h) \rangle | / _{hi} I(h)_i$, $R_{meas} = \Sigma hkl \{N(hkl)/[N(hkl) - 1]\}1/2 _i \, | Ii(hkl) - | / hkl_i \, I(hkl)$, where I is the intensity of reflection h, $_h$ is the sum over all reflections, and $_i$ is the sum over i measurements of reflection h.

3. Results and Discussion

3.1. Lipolytic Enzymes of S. chromogenes NCTC10530 and Classification of ScEst

Initially, the bacterial esterases and lipases were classified into eight families (I–VIII) and six subfamilies, all of which belong to Family I, based on the biochemical properties and sequence similarity known as the gold standard classification [17]. Recently, several newly identified lipolytic enzymes have been incorporated into the classification system, resulting in its expansion to 35 families and 11 lipase subfamilies [18].

In this study, a total of 27 putative lipolytic enzymes were identified from the in silico analysis of the genome sequence of *S. chromogenes* strain NCTC10530. These enzyme sequences were aligned with the categorized enzymes (Figure 1). Among the putative lipolytic enzymes, *Sc*Est was found to be homologous to Family XIII, specifically with thermostable carboxylesterase Est30 from *Geobacillus stearothermophilus* (AAN81911, 62.30% identity), EstOF4 from *Bacillus* sp. (AGK06467, 56.50% identity), and EstB2 from *Bacillus* sp. (AAT65181, 58.54% identity).

Multiple sequence alignment revealed that the active site of *Sc*Est shares a consensus sequence G-X-S-X-G, characteristic of the esterase/lipase family (Figure 2). *Sc*Est displayed high sequence similarity with the Family XIII proteins. However, a unique region was also identified in *Sc*Est. The amino acid sequence 103–SLNRD–107 follows the active loop in *Sc*Est in contrast to its orthologs, which have GYTVLP in the corresponding region (Figure 2). Since this site is in the vicinity of the active site, *Sc*Est may have different specificities for substrate recognition or activity. Overall, the phylogenetic and sequence analyses confirmed that *Sc*Est belongs to the XIII family but harbors a unique sequence, which may lead to a distinctive function.

3.2. Biochemical Characterization of ScEst

To confirm the esterase activity, *Sc*Est was expressed and purified using a two-step purification process. His-Tag-affinity purification followed by size-exclusion chromatography yielded the recombinant *Sc*Est protein with high purity (>95%), and a molecular weight similar to the calculated molecular weight of 29.2 kDa (Figure 3A). The molecular weight of *Sc*Est estimated by size-exclusion chromatography on FPLC was consistent with the anticipated size of the dimer (Figure 3B). The esterase activity of *Sc*Est assessed using *p*-nitrophenyl esters (*p*-NP) indicated that *Sc*Est has a substrate preference for acyl derivatives with a short chain length, and the activity declined as the size of the acyl hydrocarbon chain of the substrates increased. When the activity of *Sc*Est against *p*-nitrophenyl acetate (C2) was considered 100%, the relative activity was approximately 50% and 20% against *p*-nitrophenyl butyrate (C4) and *p*-nitrophenyl hexanoate (C6), respectively. Substrates longer than hexanoate did not show any measurable activity.

3.3. X-ray Crystallographic Study of ScEst

To determine the three-dimensional structure of *Sc*Est, crystallization screening using more than 1600 conditions, X-ray diffraction experiments, and initial model building were performed. After multiple crystallization refinements, the best single crystal was obtained with 0.25 M ammonium citrate (pH 7.0) and 20% (*w*/*v*) PEG 3350 (Figure 4A). The single crystal was cryoprotected by a brief soaking in 25% glycerol-based cryoprotectant solution and mounted under a liquid nitrogen stream at 100 K. The full coverage of 360 diffraction images was obtained at the highest resolution of 1.66 Å (Figure 4b). The space group of the *Sc*Est crystal belonged to $P2_12_12_1$ with the following unit cell parameters: a = 50.23 Å, b = 68.69 Å, c = 71.15 Å and α, β, γ = 90°. The initial structure of *Sc*Est was generated by molecular replacement using the CCP4i software suite [22]. Thermophilic carboxylesterase Est30 from *Geobacillus stearothermophilus* (PDB code, 1TQH) showed a high amino acid sequence similarity (61.79% identity) with ScEst, and was thus used as reference [23]. Model building and iterative structure refinement are currently underway using Coot software [24] and Refmac5 [25] in the CCP4i suite.

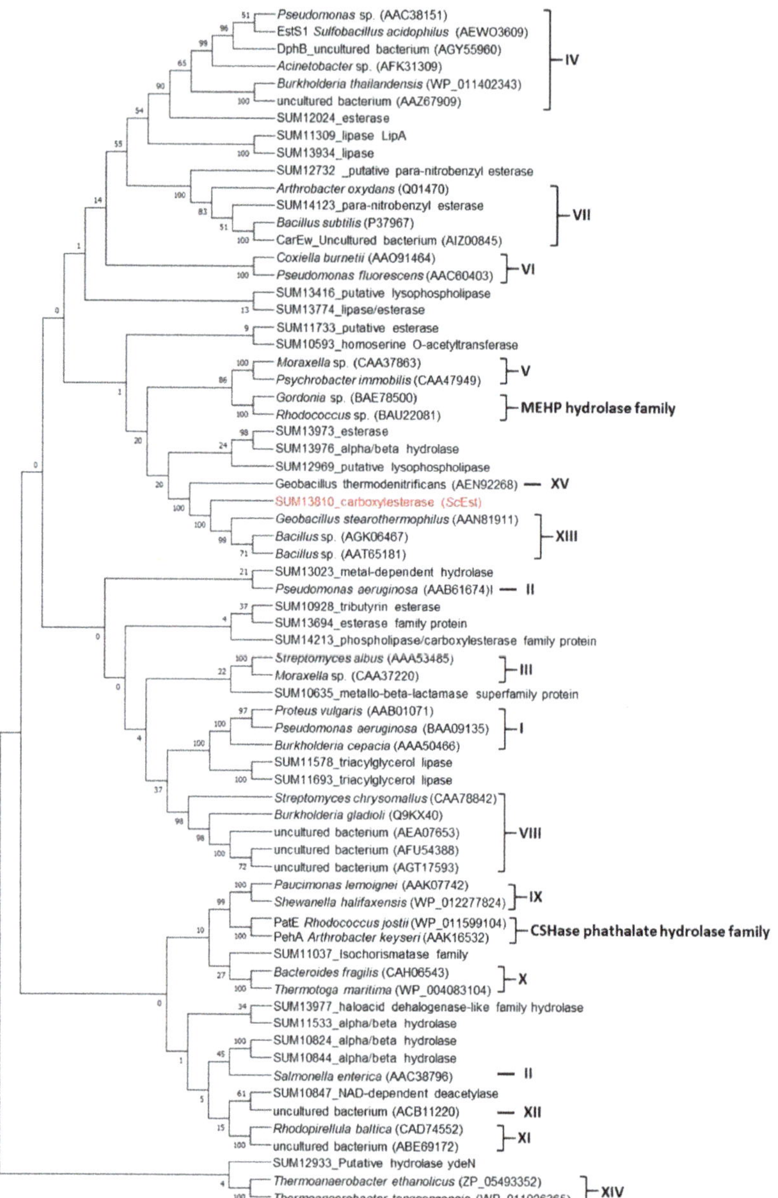

Figure 1. Phylogenetic analysis and classification of *Sc*Est with bacterial lipolytic enzyme families. Full-length protein sequences of 27 putative lipolytic enzymes from the *Staphylococcus chromogenes* strain NCTC10530 were aligned with bacterial lipolytic enzyme sequences of known categories using multiple sequence alignment (69 sequences). MEGA-X was used to create the phylogenetic tree using the neighbor-joining method. All unclear locations were deleted (using the pairwise deletion option). The percentage of duplicate trees in which the related taxa were clustered together in the bootstrap test (500 repetitions) appears next to each node. The GenBank accession numbers are indicated in parentheses.

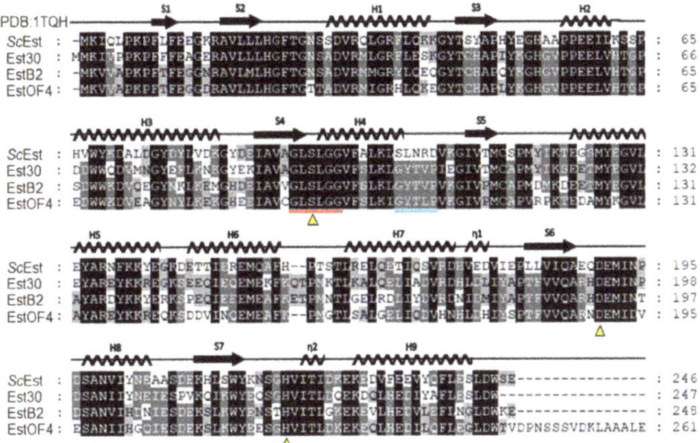

Figure 2. Multiple sequence alignment of *Sc*Est with other esterases of Family VIII. The sequences including that of thermostable carboxylesterase Est30 from *Geobacillus stearothermophilus* (GenBank AAN81911), EstOF4 from *Bacillus pseudofirmus* (GenBank AGK06467), and EstB2 from *Bacillus* sp. 01-855 (Genbank AAT65181) belonging to the bacterial lipolytic enzyme Family VIII were aligned using ClustalX. The conserved sites are highlighted in a darker color, whereas varied or polymorphic sites are shown in a lighter color. The secondary structure deduced from the Est30 structure (PDB code 1TQH) is displayed on the top of the aligned sequences. The conserved sequence at the active site characteristic of Family VIII is indicated with a red bar. The adjacent region specific to the *Sc*Est is marked with a cyan bar. The conserved catalytic triads are indicated with triangles.

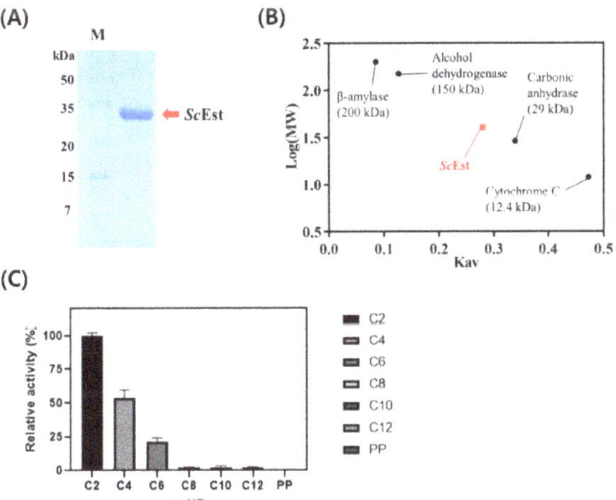

Figure 3. Purification and characterization of *Sc*Est. (**A**) SDS-PAGE of purified *Sc*Est along with a molecular weight marker. (**B**) Size-exclusion chromatography (SEC) of *Sc*Est. The elution time of *Sc*Est was integrated with the calibration curve obtained using molecular weight standards β-amylase (200 kDa), alcohol dehydrogenase (150 kDa), carbonic anhydrase (29 kDa), and cytochrome C (12 kDa). Kav = (Vs − Vo)/(Vc − Vo). vs. = elution time; Vo: column void volume; Vc: column volume. (**C**) Evaluation of esterase activity of *Sc*Est using 1 mM *p*-Nitrophenyl esters as substrates in 50 mM sodium phosphate buffer at pH 7.0. *p*-Nitrophenyl esters used in the activity assay were C2, *p*-Nitrophenyl acetate; C4, *p*-Nitrophenyl butyrate; C6, 4-Nitrophenyl hexanoate; C8, *p*-Nitrophenyl octanoate; C10, *p*-Nitrophenyl decanoate; C12, *p*-Nitrophenyl dodecanoate; PP, phenyl palmitate. The activity of recombinant *Sc*Est against *p*-NA(C2) is represented as 100%, whereas the relative activities against other substrates are shown in percentage.

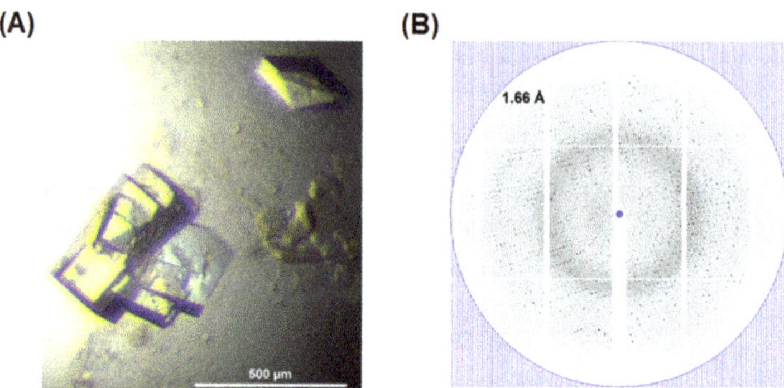

Figure 4. Preliminary X-ray crystallographic study of *Sc*Est. (**A**) *Sc*Est crystals for diffraction experiment obtained in 0.25 M ammonium citrate tribasic (pH 7.0), 20% PEG 3350. (**B**) Diffraction image of *Sc*Est crystal with the highest resolution value in the last atomic shell (1.69–1.66 Å). Blue circle represents the highest resolution range, and diffraction spots are shown at a resolution of 1.66 Å.

4. Conclusions

The biochemical characteristics of a carboxylesterase *Sc*Est, derived from *S. chromogenes* NCTC10530, which is the most common bacterial pathogen causing infectious diseases in dairy cows, were examined. The *ScEst* gene was identified, isolated, overexpressed in *E. coli*, and the protein was purified with affinity columns and size-exclusion chromatography. The *Sc*Est enzyme prefers the acyl derivatives with a short chain length as substrates. A preliminary crystallographic investigation of *Sc*Est resulted in a high-resolution dataset. We anticipate that elaborating the structure-based enzymatic mechanism of *Sc*Est will provide valuable information for understanding pathogenic esterases and designing ester prodrugs to treat MDR bacteria.

Author Contributions: Conceptualization, J.H.L. and T.D.K.; Methodology, J.H., W.Y. and S.J.; Validation, J.H., S.J., J.C. and K.K.K.; Formal Analysis, H.D. and M.J.L.; Writing—Original Draft Preparation, H.D., J.H. and T.D.K.; Writing—Review & Editing, H.D. and T.D.K.; Visualization, J.H. and S.J.; Supervision, H.D. and T.D.K.; Funding Acquisition, J.H.L. and T.D.K. All authors have read and agreed to the published version of the manuscript.

Funding: This research was a part of the project titled "Development of potential antibiotic compounds using polar organism resources (15250103, KOPRI Grant PM22030)" funded by the Ministry of Oceans and Fisheries, Republic of Korea. This research was also supported by a National Research Foundation of Korea Grant from the Korean Government (MSIT; the Ministry of Science and ICT) (NRF-2021M1A5A1075524) (KOPRI-PN22014), and an academic grant (NRF-2021R1F1A1048135) from the National Research Foundation of Korea (T.D.K).

Institutional Review Board Statement: Not applicable.

Informed Consent Statement: Not applicable.

Data Availability Statement: Not applicable.

Acknowledgments: We would like to thank the staff at the X-ray core facility of the Korea Basic Science Institute (KBSI; Ochang, Korea) and BL-7A of the Pohang Accelerator Laboratory (Pohang, Korea) for their kind help with data collection.

Conflicts of Interest: The authors declare no conflict of interest.

References

1. Nikaido, H. Multidrug resistance in bacteria. *Annu. Rev. Biochem.* **2009**, *78*, 119–146. [CrossRef] [PubMed]
2. Serra-Burriel, M.; Keys, M.; Campillo-Artero, C.; Agodi, A.; Barchitta, M.; Gikas, A.; Palos, C.; López-Casasnovas, G. Impact of multi-drug resistant bacteria on economic and clinical outcomes of healthcare-associated infections in adults: Systematic review and meta-analysis. *PLoS ONE* **2020**, *15*, e0227139. [CrossRef] [PubMed]

3. Zárate, S.G.; De La Cruz Claure, M.L.; Benito-Arenas, R.; Revuelta, J.; Santana, A.G.; Bastida, A. Overcoming aminoglycoside enzymatic resistance: Design of novel antibiotics and inhibitors. *Molecules* **2018**, *23*, 284. [CrossRef] [PubMed]
4. Fair, R.J.; Tor, Y. Antibiotics and bacterial resistance in the 21st century. *Perspect. Medicin. Chem.* **2014**, *6*, 25–64. [CrossRef] [PubMed]
5. Jubeh, R.; Breijyeh, B.; Karaman, Z. Antibacterial prodrugs to overcome bacterial resistance. *Molecules* **2020**, *25*, 1543. [CrossRef]
6. Sakagami, K.; Atsumi, K.; Tamura, A.; Yoshida, T.; Nishihata, K.; Fukatsu, S.J. *Antibiot.* Synthesis and oral activity of ME 1207, a new orally active cephalosporin. *J. Antibiot.* **1990**, *8*, 8–11. [CrossRef]
7. Maag, H. *Prodrugs of Carboxylic Acids, Prodrugs*; Springer: New York, NY, USA, 2008; pp. 703–729. [CrossRef]
8. Hamada, Y. Recent progress in prodrug design strategies based on generally applicable modifications. *Bioorg. Med. Chem. Lett.* **2017**, *27*, 1627–1632. [CrossRef]
9. Elayyan, S.; Karaman, D.; Mecca, G.; Scrano, L.; Bufo, S.A.; Karaman, R. Antibacterial predrugs-from 1899 till 2015. *World J. Pharm. Pharm. Sci.* **2015**, *4*, 1504–1529. [CrossRef]
10. Larsen, E.M.; Johnson, R.J. Microbial esterases and ester prodrugs: An unlikely marriage for combating antibiotic resistance. *Drug Dev. Res.* **2019**, *80*, 33–47. [CrossRef]
11. Satoh, T.; Hosokawa, M. Structure, function and regulation of carboxylesterases. *Chem. Biol. Interact.* **2006**, *162*, 195–211. [CrossRef]
12. Mikati, M.O.; Miller, J.J.; Osbourn, D.M.; Barekatain, Y.; Ghebremichael, N.; Shah, I.T.; Burnham, C.D.; Heidel, K.M.; Yan, V.C.; Muller, F.L.; et al. Antimicrobial prodrug activation by the Staphylococcal glyoxalase GloB. *ACS Infect. Dis.* **2020**, *6*, 3064–3075. [CrossRef] [PubMed]
13. Wang, Y.; Le, L.T.H.L.; Yoo, W.; Lee, C.W.; Kim, K.K.; Lee, J.H.; Kim, T.D. Characterization, immobilization, and mutagenesis of a novel cold-active acetylesterase (EaAcE) from Exiguobacterium antarcticum B7. *Int. J. Biol. Macromol.* **2019**, *136*, 1042–1051. [CrossRef] [PubMed]
14. Oh, C.; Ryu, B.H.; An, D.R.; Nguyen, D.D.; Yoo, W.; Kim, T.; Ngo, T.D.; Kim, H.S.; Kim, K.K.; Kim, T.D. Structural and biochemical characterization of an octameric carbohydrate acetylesterase from *Sinorhizobium meliloti*. *FEBS Lett.* **2016**, *590*, 1242–1252. [CrossRef] [PubMed]
15. Lee, C.W.; Kwon, S.; Park, S.H.; Kim, B.Y.; Yoo, W.; Ryu, B.H.; Kim, H.W.; Shin, S.C.; Kim, S.; Park, H.; et al. Crystal structure and functional characterization of an esterase (EaEST) from exiguobacterium antarcticum. *PLoS ONE* **2017**, *12*, e0169540. [CrossRef] [PubMed]
16. Sarkar, J.; Dutta, A.; Pal Chowdhury, P.; Chakraborty, J.; Dutta, T.K. Characterization of a novel family VIII esterase EstM2 from soil metagenome capable of hydrolyzing estrogenic phthalates. *Microb. Cell Fact.* **2020**, *19*, 77. [CrossRef] [PubMed]
17. Arpigny, J.L.; Jaeger, K.E. Bacterial lipolytic enzymes: Classification and properties. *Biochem. J.* **1999**, *343*, 177–183. [CrossRef]
18. Hitch, T.C.A.; Clavel, T. A proposed update for the classification and description of bacterial lipolytic enzymes. *PeerJ* **2019**, *7*, e7249. [CrossRef]
19. Larkin, M.A.; Blackshields, G.; Brown, N.P.; Chenna, R.; Mcgettigan, P.A.; McWilliam, H.; Valentin, F.; Wallace, I.M.; Wilm, A.; Lopez, R.; et al. Clustal W and Clustal X version 2.0. *Bioinformatics* **2007**, *23*, 2947–2948. [CrossRef]
20. Kumar, S.; Stecher, G.; Li, M.; Knyaz, C.; Tamura, K. MEGA X: Molecular evolutionary genetics analysis across computing platforms. *Mol. Biol. Evol.* **2018**, *35*, 1547–1549. [CrossRef] [PubMed]
21. Peng, Y.; Fu, S.; Liu, H.; Lucia, L.A. Accurately determining esterase activity via the isosbestic point of p-nitrophenol. *BioResources* **2016**, *11*, 10099–10111. [CrossRef]
22. Vagin, A.; Teplyakov, A. Molecular replacement with MOLREP. *Acta Crystallogr. D Biol. Crystallogr.* **2010**, *66*, 22–25. [CrossRef]
23. Liu, P.; Wang, Y.F.; Ewis, H.E.; Abdelal, A.T.; Lu, C.D.; Harrison, R.W.; Weber, I.T. Covalent reaction intermediate revealed in crystal structure of the Geobacillus stearothermophilus carboxylesterase Est30. *J. Mol. Biol.* **2004**, *342*, 551–561. [CrossRef]
24. Winn, M.D.; Ballard, C.C.; Cowtan, K.D.; Dodson, E.J.; Emsley, P.; Evans, P.R.; Keegan, R.M.; Krissinel, E.B.; Leslie, A.G.W.; McCoy, A.; et al. Overview of the CCP4 suite and current developments. *Acta Crystallogr. D Biol. Crystallogr.* **2011**, *67*, 235–242. [CrossRef]
25. Murshudov, G.N.; Skubák, P.; Lebedev, A.A.; Pannu, N.S.; Steiner, R.A.; Nicholls, R.A.; Winn, M.D.; Long, F.; Vagin, A.A. REFMAC5 for the refinement of macromolecular crystal structures. *Acta Crystallogr. D Biol. Crystallogr.* **2011**, *67*, 355–367. [CrossRef]

Communication

Identification, Characterization, and Preliminary X-ray Diffraction Analysis of a Single Stranded DNA Binding Protein (LjSSB) from Psychrophilic *Lacinutrix jangbogonensis* PAMC 27137

Woong Choi [1,†], Jonghyeon Son [2,†], Aekyung Park [3], Hongshi Jin [4], Seung Chul Shin [4], Jun Hyuck Lee [1,5], T. Doohun Kim [6] and Han-Woo Kim [1,5,*]

1. Research Unit of Cryogenic Novel Material, Korea Polar Research Institute, Incheon 21990, Korea; woong@kopri.re.kr (W.C.); junhyucklee@kopri.re.kr (J.H.L.)
2. New Drug Development Center, Daegu-Gyeongbuk Medical Innovation Foundation, Daegu 41061, Korea; tuutoo92@dgmif.re.kr
3. Division of Emerging Infectious Diseases Bureau of Infectious Disease Diagnosis Control, Korea Disease Control and Prevention Agency, Cheongju 28159, Korea; parkak1003@gmail.com
4. Division of Life Sciences, Korea Polar Research Institute, Incheon 21990, Korea; zoeykim@kopri.re.kr (H.J.); ssc@kopri.re.kr (S.C.S.)
5. Department of Polar Sciences, University of Science and Technology, Incheon 21990, Korea
6. Department of Chemistry, Sookmyung Women's University, Seoul 04310, Korea; doohunkim@sookmyung.ac.kr

* Correspondence: hwkim@kopri.re.kr

† These authors contributed equally to this work.

Citation: Choi, W.; Son, J.; Park, A.; Jin, H.; Shin, S.C.; Lee, J.H.; Kim, T.D.; Kim, H.-W. Identification, Characterization, and Preliminary X-ray Diffraction Analysis of a Single Stranded DNA Binding Protein (LjSSB) from Psychrophilic *Lacinutrix jangbogonensis* PAMC 27137. *Crystals* **2022**, *12*, 538. https://doi.org/10.3390/cryst12040538

Academic Editor: Abel Moreno

Received: 23 March 2022
Accepted: 8 April 2022
Published: 11 April 2022

Abstract: Single-stranded DNA-binding proteins (SSBs) are essential for DNA metabolism, including repair and replication, in all organisms. SSBs have potential applications in molecular biology and in analytical methods. In this study, for the first time, we purified, structurally characterized, and analyzed psychrophilic SSB (LjSSB) from *Lacinutrix jangbogonensis* PAMC 27137 isolated from the Antarctic region. LjSSB has a relatively short amino acid sequence, consisting of 111 residues, with a molecular mass of 12.6 kDa. LjSSB protein was overexpressed in *Escherichia coli* BL21 (DE3) and analyzed for binding affinity using 20- and 35-mer deoxythymidine oligonucleotides (dT). In addition, the crystal structure of LjSSB at a resolution 2.6 Å was obtained. The LjSSB protein crystal belongs to the space group C222 with the unit cell parameters of a = 106.58 Å, b = 234.14 Å, c = 66.14 Å. The crystal structure was solved using molecular replacement, and subsequent iterative structure refinements and model building are currently under progress. Further, the complete structural information of LjSSB will provide a novel strategy for protein engineering and for the application on molecular biological techniques.

Keywords: single stranded DNA binding protein; *Lacinutrix jangbogonensis* PAMC 27137; X-ray crystallography

1. Introduction

For important cellular processes, chromosomal DNA must be in a single-stranded form. Therefore, single-stranded DNA (ssDNA) intermediates are created by DNA unwinding and serve as templates for myriad cellular functions [1]. However, ssDNA is less stable and more sensitive to chemicals than double-stranded DNA (dsDNA); thus, ssDNA is more prone to damage than dsDNA. DNA lesions can interfere with essential cellular processes, ultimately affecting cell viability. These problems can be solved by encoding specialized ssDNA-binding proteins (SSBs) [2]. SSBs bind to ssDNA with high affinity in a sequence-independent manner to protect it from chemical and enzymatic damage [3]. SSBs are present in all living organisms and play essential roles in many processes related to DNA metabolism, such as DNA replication, repair, and homologous genetic recombination [4]. Most bacterial SSBs comprise two domains: an N-terminal domain called oligosaccharide/oligonucleotide binding (OB) domain, composed of a minimum of five β-sheets arranged as a β barrel capped by a single α helix and responsible for both ssDNA binding and oligomerization, and a C-terminal domain, that involves protein-protein interaction [5–7]. The OB domain is separated from the highly conserved last nine amino acids of the C-terminal domain by a proline-or glycine-rich linker, called intrinsically disordered linker (IDL) [8]. The C-terminal domain of SSB consists of IDL and last nine highly conserved residues, called acidic tip, and interacts with various other proteins involved in cell survival [9,10]. Although SSBs are found in every organism, they have little sequence similarity and differ in their subunit composition and oligomerization state [11]. While most bacterial SSBs form a homotetramer and have a single OB fold per polypeptide, eukaryotic SSBs, commonly called replication protein A (RPA), generally function as heterotrimeric complexes that contain six OB folds distributed among three subunits [12–14].

Due to these properties, there has been growing interest in SSBs owing to their potential applications in molecular biology and analytical methods. An increasing number of studies have found that SSBs increase amplification efficiency. Previous reports have shown that PCR application of an ssDNA-binding protein isolated from bacteriophage T4-infected *Escherichia coli*, named gene 32 protein, results in an increased amplification efficiency [15,16]. In particular, thermostable SSBs can be used in applications that require extremely high temperatures, such as nucleic acid amplification and sequencing [17]. Furthermore, the affinity of SSB towards ssDNA has been successfully utilized for rolling circle amplification (RCA) [18]. The addition of SSBs during the RCA reaction prevents the generation of double-stranded DNA, and consistently produces single-stranded products without termination. Although mesophilic and thermophilic SSBs have been identified and extensively studied, there is little information about psychrophilic SSBs. Cold-adapted enzyme proteins are useful for molecular biosciences due to the need for enzymes to be used in sequential reactions, and to be inactivated after performing their individual function. Heat-labile enzymes enable heat inactivation at temperatures that do not cause double-stranded DNA to melt, and their use obviates the need to use chemical extraction processes [19]. Particularly, molecular biological techniques, such as isothermal amplification or whole genome amplification, require enzymes that have activity at a relatively low temperature because these techniques have to be performed within room temperature range. Therefore, the characterization of cold-active SSBs offers an attractive alternative to other thermostable SSBs in molecular biology applications.

In this study, we identified and purified a psychrophilic SSB protein, called LjSSB, from a *Lacinutrix jangbogonensis* PAMC 27137 Gram-negative strain isolated from an Antarctic region [20]. We investigated whether LjSSB has a different binding affinity pattern depending on ssDNA length. Furthermore, LjSSB was crystallized for X-ray diffraction experiments to obtain structural information. We successfully determined LjSSB structure, and further iterative refinement and model building are currently in progress. We believe that our biochemical and structural analyses will help prospective enzyme engineering for applications in molecular biology.

2. Materials and Methods

2.1. Sequence Analysis for LjSSB

To investigate the properties of the amino acid sequence of LjSSB, the sequence was compared with those of several previously classified SSBs. A total of 32 amino acid sequences containing LjSSB were used for multiple sequence alignment, and a phylogenetic tree was generated based on the neighbor-joining method using ClustalW [21] and MEGA-X [22].

2.2. Cloning, Expression, and Purification of the LjSSB from L. jangbogonensis

Genomic DNA of PAMC 27,137 was isolated using a DNA purification kit according to the manufacturer's instructions (Promega, Madison, WI, USA). For cloning into the plasmid, the LjSSB gene was amplified from genomic DNA by PCR. The PCR product was purified and cloned into the pET22b(+) vector, which has a C-terminal hexa-histidine tag between the NdeI and XhoI restriction sites. Subsequently, the cloned sequence was estimated by Sanger sequencing using a T7 promoter and terminator primer pair. For protein expression, plasmids harboring LjSSB were transformed into *E. coli* BL21 (DE3), and transformed cells were cultured in Luria–Bertani (LB) broth. When OD_{600} reached 0.6–0.8, cells were induced by 0.3 mM IPTG and incubated at 25 °C for 16 h. Subsequently, cells were harvested by centrifugation at 8000× *g* for 20 min at 4 °C. For purification of SSB proteins, pelleted cells were resuspended in lysis buffer containing 20 mM Tris-HCl pH 8.0, and 200 mM NaCl and lysed by sonication. The supernatant was subjected to a column charged with a Ni^{2+}-chelated resin. The column was washed with washing buffer (20 mM Tris-HCl, pH 8.0, 200 mM NaCl, and 30 mM imidazole) and the protein was eluted by elution buffer (20 mM Tris-HCl pH 8.0, 200 mM NaCl, and 300 mM imidazole). This was followed by further purification via gel filtration chromatography using a HiLoad 16/60 Superdex 200 column (GE Healthcare, Chicago, IL, USA) pre-equilibrated by buffer with same composition of lysis buffer.

2.3. Estimation of Oligomerization State

The oligomerization state of LjSSB was determined using analytical size-exclusion chromatography. The molecular mass of recombinant SSB was measured by elution analysis of standard proteins from a Superdex 200 10/300 GL column (GE Healthcare, IL, USA). The column was calibrated using molecular mass standards: thyroglobulin (640 kDa), γ-globulin (155 kDa), ovalbumin (47 kDa), and ribonuclease A (13.7 kDa).

2.4. Enzymatic Analysis Using Gel Electrophoresis Mobility Shift Assay (EMSA)

Deoxythymidine (dT) oligonucleotides (20- and 35-mer) were synthesized by Macrogen (Seoul, Korea). LjSSB and ssDNA binding reaction was performed in 20 μL volumes containing each various concentrations of LjSSB (0.125, 0.25, 0.5, 1, 2, 4, 8, 16, and 32 μM) with 500 nM of synthesized dT20 and dT35 in a binding buffer (20 mM Tris—HCl pH 8.0, and 200 mM NaCl). The reaction mixtures were incubated for 15 min at room temperature (25 °C). Subsequently, the reaction products with oligonucleotides were loaded onto 8% acrylamide gel and separated by electrophoresis in a 1X TBE buffer (89 mM Tris borate, pH 8.0, and 2 mM EDTA). The gel was stained with SYBR Gold (Thermo Fisher Scientific, Waltham, MA, USA) for 15 min at room temperature (RT, at 25 °C). Bands corresponding to unbound ssDNA and SSB ssDNA complex were visualized under UV light and photographed.

2.5. Crystallization, Data Collection, and Refinement for LjSSB

The crystallization of LjSSB was performed using the following procedure: the LjSSB was concentrated to 20 mg/mL using a 3 kDa molecular weight cutoff spin concentrator (Millipore, Burlington, MA, USA). Initial crystallization screening of LjSSB was performed by the sitting-drop vapor-diffusion method, using commercially available screening solution kits: MCSG I to IV (Anatrace Inc., Maumee, OH, USA), SaltRx HT™ (Hampton Research Corp., Aliso Vieojo, CA, USA), and JCSG™ (Molecular Dimensions Inc., Altamonte Springs, FL, USA). Each crystallization drop was prepared by equilibrating the mixture of 0.5 μL reservoir solution and 0.5 μL protein solution against 80 μL reservoir solution using a mosquito® liquid-handling robot (TTP Labtech Ltd., Melbourn, Hertfordshire, UK). Diffraction data for LjSSB were obtained under conditions comprising 0.1 M of phosphate-citrate pH 4.2 and 1.6 M of NaH_2PO_4/0.4M K_2HPO_4. Subsequently, crystals were transferred to perfluoropolyether oil PFO-X175/08 (Hampton Research Corp., CA, USA) to protect cryodamage and flash-frozen in a liquid nitrogen stream. A single crystal was mounted on the goniometer sample holder. A total of 360 diffraction image data of LjSSB were collected using an ADSC Quantum 315 CCD detector on beamline 5C at the Pohang Accelerator Laboratory (PAL, Pohang, Korea) by rotating with 1° of oscillation per frame. X-ray diffraction data were collected at a resolution of 2.6 Å. The datasets were indexed, processed, and scaled using the XDS program [23]. The initial crystal structure of LjSSB was determined using the molecular replacement (MR) method with a single-stranded DNA binding protein from *Escherichia coli* (EcSSB, PDB code: 4MZ9) as a search model. The collection data are summarized in Table 1.

Table 1. Initial X-ray diffraction data.

Data Collection	
Wavelength (Å)	0.9796
X-ray source	PAL 5C
Rotation range per image (°)	1
Exposure Time (s)	0.5
Space group	C 2 2 2
Unit-cell parameters (Å, °)	a = 106.58, b = 231.14, c = 66.14
Resolution range (Å) [a]	50.00–2.60 (2.70–2.60)
No. of observed reflections	71,319
No. of unique reflections [a]	37,560 (3753)
Completeness (%) [a]	99.4 (98.7)
Redundancy [a]	7.1 (7.6)
R_{sym} [a,b]	0.138 (0.771)
R_{meas} [a,b]	0.144 (0.800)
I/σ [a]	14.23 (4.67)
CC(1/2) [c] (%)	99.9 (94.6)

[a] Values in parentheses correspond to the highest-resolution shells. [b] Rmerge = $\Sigma h \Sigma i \mid I(h)_i - \langle I(h) \rangle \mid / \Sigma h \Sigma i\, I(h)$, where I is the intensity of reflection h, Σh is the sum over all reflections, and Σi is the sum over i measurements of reflection h. [c] Percentage correlation between intensities from random half datasets.

3. Results and Discussion

3.1. Sequence Analysis for LjSSB

SSB consists of two domains: an N-terminal OB-domain and a C-terminal domain. The N-terminal domain of SSB is involved in ssDNA binding, and the C-terminal domain is composed of IDL and an acidic tip and is involved in protein-protein interactions, as with DNA repair or recombination proteins [24]. The results of genomic analysis of *L. jangbogonensis* PAMC 27137 (GenBank accession No. JSWF00000000) [20] indicated that they have three putative SSB genes. We attempted to clone and express all three; however, two SSBs were expressed in insoluble form, and only one SSB was soluble and expressed as recombinant proteins in *E. coli*. The soluble SSB, called LjSSB, and the LjSSB predicted protein contained 111 amino acid residues. Among the known bacterial SSB proteins, LjSSB has the smallest molecular size.

Comparative protein sequence analysis revealed that LjSSB has an N-terminal OB fold (Figure 1A) but not a C-terminal domain. Although the results of multiple sequence alignment displayed overall sequence identities of 45% to *E. coli* (EcSSB, GenBank ID: EEV5779109.1), 45% to *Enterobacter cloacae* (EncSSB, GenBank ID: WP_094085120.1), 36% to *Bacillus subtilis* (BsuSSB, GenBank ID: WP_153257179.1), and 33% to *Thermotoga maritima* (TmaSSB, GenBank ID: WP_004081225.1), they also showed that specific residues involved in binding ssDNA were highly conserved (Figure 1B). However, oligomerization-related residues were less conserved than ssDNA binding sites.

Multiple sequence alignments for evolutionary analysis are classified into three groups based on their origin: psychrophilic, mesophilic, and thermophilic SSB. The phylogenetic tree showed that LjSSB was part of the psychrophilic SSB group (Figure 1C).

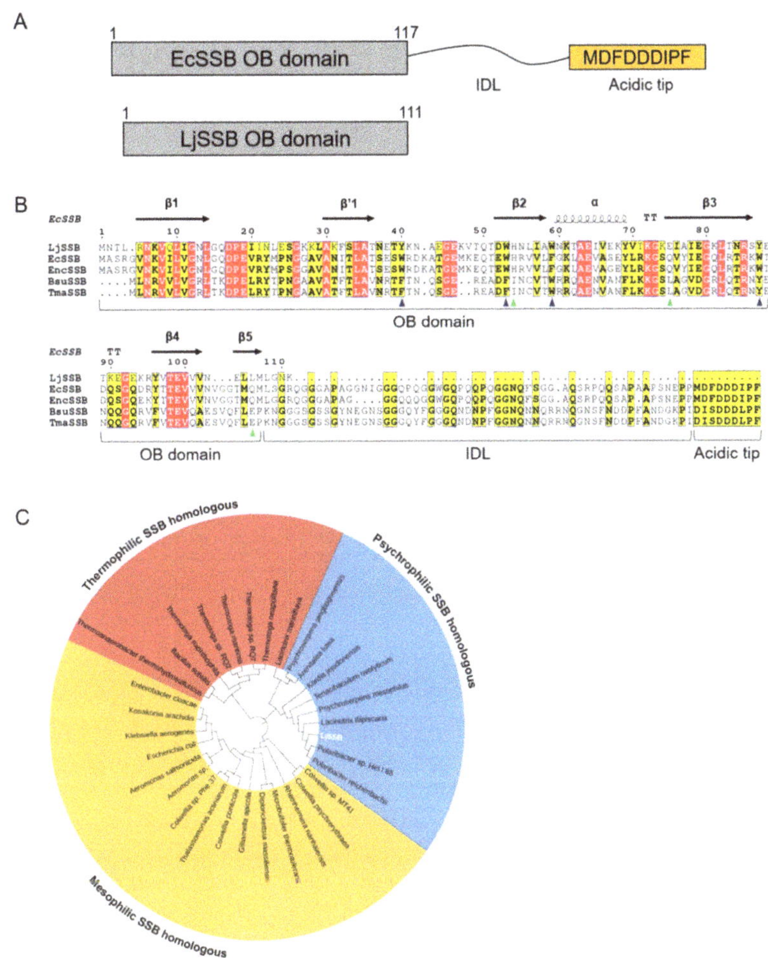

Figure 1. Sequence analysis for LjSSB. (**A**) The comparison of domain architecture of LjSSB and EcSSB. In contrast with EcSSB, LjSSB does not have a C-terminal domain. (**B**) Multiple sequence alignment for SSBs. LjSSB from *Lacinutrix jangbogonensis* PAMC 27,137 in this study; EcSSB from *Escherichia coli* (GenBank ID: EEV5779109.1); EncSSB from *Enterobacter cloacae* (GenBank ID: WP_094085120.1); BsuSSB from *Bacillus subtilis* (GenBank ID: WP_153257179.1); and TmaSSB from *Thermotoga maritima* (GenBank ID: WP_004081225.1). The secondary structure deduced from EcSSB (PDB code: 1KAW) is showed on the top of aligned sequences. The strictly conserved regions are highlighted in red, and homologous regions are highlighted in yellow. The ssDNA binding site and oligomerization-involved residues from EcSSB are labeled using blue and green triangles, respectively. (**C**) The amino acid sequences for psychrophilic, mesophilic, and thermophilic SSBs were aligned using ClustalW, followed by construction of neighbor-joining (NJ) phylogenetic trees with 1000 bootstrap replicates using MEGA-X software. The tree display was obtained with online iTOL [25].

3.2. Oligomerization Status for LjSSB

SSB proteins have a variety of oligomeric states, ranging from monomers to pentamers [26–29]. In the case of bacterial SSBs, mesophilic SSB function as homotetramers with a single OB-domain subunit. Thermophilic SSBs from bacteria, such as *Deinococcus radiodurans* and *Thermus aquaticus*, exist as homodimers. Analytical gel filtration was performed to determine the oligomeric status of the proteins in the solution. The elution profile relative to standard proteins indicated that the native molecular mass of LjSSB was approximately 50 kDa, which is nearly four times the molecular mass of the LjSSB monomer (12.6 kDa) (Figure 2). Since the molecular mass of LjSSB is 50.4 kDa as calculated from its amino acid composition in the case of homotetramers, the result implied that LjSSB existed as a stable homotetramer in the solution.

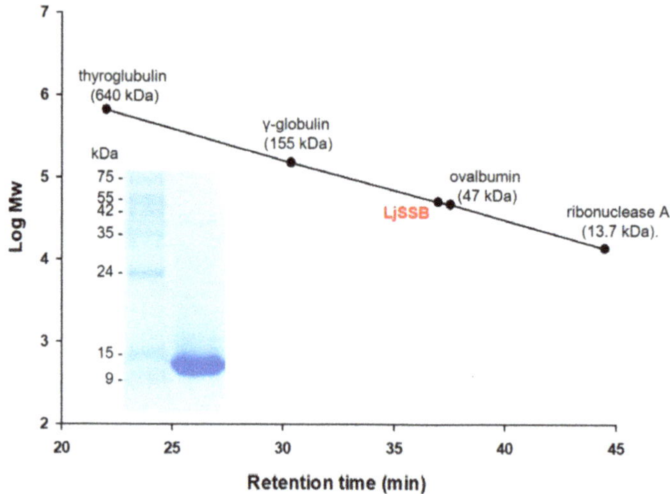

Figure 2. Oligomerization state for LjSSB. Purified LjSSB protein in buffer A (20 mM Tris-HCl pH 8, 200 mM NaCl) was applied to a Superdex 200 10/300 GL column equilibrated with the same buffer. The LjSSB was detected at 280 nm; 12% of Coomassie Blue-stained SDS-PAGE for the purified LjSSB and molecular mass standards are also shown. The molecular mass of purified LjSSB is 12.6 kDa.

3.3. Binding Properties for LjSSB with ssDNA

Most bacterial SSBs have different binding modes for ssDNA of various lengths [7]. To investigate the binding mode and affinity of LjSSB for ssDNA, we performed gel shift assay (EMSA) using different concentrations of purified LjSSB (0–32 μM) and 20- and 35-mer oligonucleotides consisting of deoxythymidine (dT20 and dT35, respectively). As shown in Figure 3A, when LjSSB was incubated with dT20, intensity of the unbound DNA band decreased, but no significant band shift was observed at a low protein concentration (lanes 3 to 5 in Figure 3A). This result suggested that a low concentration of LjSSB may not form a stable complex with dT20. Nevertheless, at an LjSSB concentration of 1 μM (lane 7 in Figure 3A), a significant band shift was observed, indicating that LjSSB formed a single complex with dT20. Furthermore, when the LjSSB concentration was increased (lanes 8–10 in Figure 3A), another slow-migrating band appeared. This slow migrated second band indicated that LjSSB formed a second complex with dT20 under a high concentration of protein. At least two LjSSB tetramers may be involved per oligonucleotide, despite the short length of dT20. This phenomenon was also observed in LjSSB incubated with dT35 (Figure 3B). At low concentrations of LjSSB, 0.5 to 2 μM (lanes 4 to 6 in Figure 3B), a single complex was observed, whereas a second complex was detected under high concentrations of protein (lanes 8 to 10 in Figure 3B). The EMSA results for dT20 and dT35 with LjSSB indicated that LjSSB forms two types of complexes with dT20 and dT35 according to protein concentration, which was a distinctive feature from other SSBs, that had C-termini and formed only a single complex with short ssDNA [30,31]. When other SSBs with C-terminal domains were incubated with short ssDNA (15–40 nt), only a single complex was detected, contrary to the observed results for LjSSB. These different features of LjSSB are considered to derive from the absence of a C-terminal domain because the IDL and acidic tip of the C-terminal domain are involved in mediating ssDNA binding [32,33].

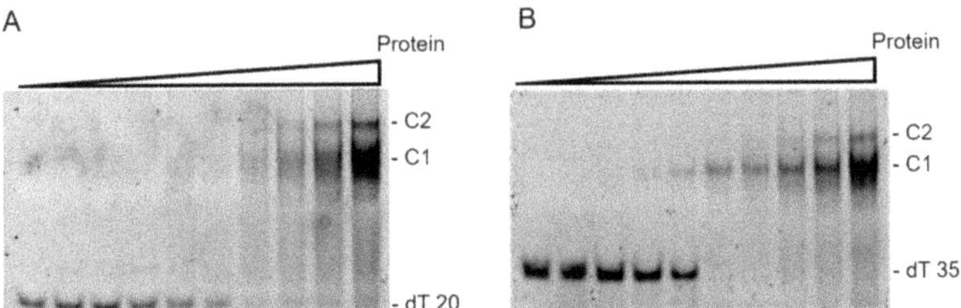

Figure 3. Gel shift assay for Binding of LjSSB with dT20 and dT35. LjSSB (0, 0.125, 0.25, 0.5, 1, 2, 4, 8, 16, and 32 μM) was incubated with 500 nM of dT oligomers. (**A**) dT20 and (**B**) dT35 incubated with LjSSBs. C1 and C2 refer to complex 1 and complex 2, respectively.

3.4. *Crystallization and X-ray Structure Analysis of LjSSB*

To obtain structural information on LjSSB, crystallization screening and X-ray diffraction experiments were performed. The crystallization conditions of LjSSB were screened using more than 1000 different crystallization buffers. The best single crystal was obtained under 0.1 M of phosphate-citrate pH 4.2 and 1.6 M of NaH_2PO_4/0.4 M K_2HPO_4 (Figure 4A). To prevent freezing damage, the single crystals were soaked in a cryoprotectant, perfluoropolyether oil PFO-X175/08. Subsequently, X-ray experiments were performed by mounting the samples under a liquid nitrogen stream. The 360 frames of the diffraction images were collected at a resolution of approximately 2.6 Å (Figure 4B). The acquired diffraction data were indexed, integrated, and scaled using the XDS program [23]. The crystal belongs to the orthorhombic space group C222, with the unit cell parameters a = 106.58 Å, b = 234.14 Å, c = 66.14 Å. Molecular replacement was performed for phase determination using MOLREP [34] and Phaser [35]. The initial structure of LjSSB was determined using EcSSB (PDB code: 1KAW) as a reference model. Refinement and model building are currently underway using the Phenix [36] and COOT software programs [37].

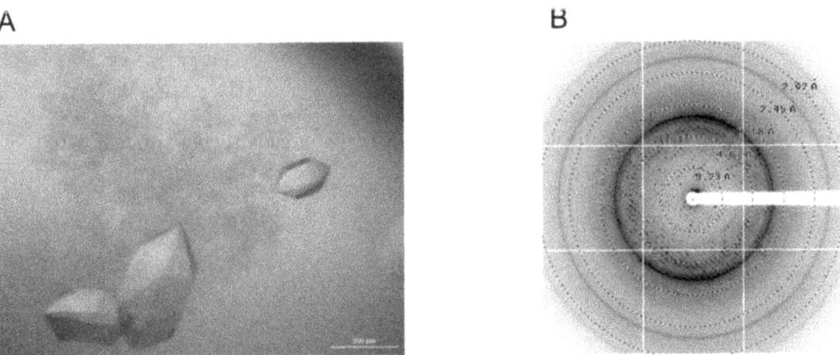

Figure 4. Crystal and preliminary X-ray diffraction analysis of LjSSB. (**A**) The LjSSB crystal obtained under 0.1 M of phosphate-citrate pH 4.2 and 1.6 M of NaH_2PO_4/0.4 M K2HPO4. (**B**) The diffraction image of LjSSB crystal. The diffraction spots are indicated at a resolution of 2.6 Å.

4. Conclusions

In this study, we identified the biochemical properties of a psychrophilic SSB, LjSSB, from *L. jangbogonensis* PAMC 27137. The LjSSB gene was identified, expressed in *E. coli*, and the resulting protein was purified using Ni^+ affinity column and gel filtration chromatography. Sequence alignment and PAGE results indicated that LjSSB is the smallest known bacterial SSB protein because it has only an N-terminal domain. In addition, LjSSB has a different pattern of binding affinity according to ssDNA length. Structural refinement and model building are currently in progress. We believe that further structural analysis of LjSSB will provide new insights for engineering SSBs for application in molecular biology techniques.

Author Contributions: Conceptualization, W.C. and H.-W.K.; methodology, W.C., H.J. and A.P.; validation, W.C. and J.S.; formal analysis, W.C., H.J., S.C.S. and A.P.; writing—original draft preparation, W.C. and H.-W.K.; writing—review and editing, W.C., S.C.S., J.H.L., T.D.K. and H.-W.K., visualization, W.C. and J.S.; supervision, H.-W.K.; funding acquisition, H.-W.K. All authors have read and agreed to the published version of the manuscript.

Funding: This research was supported by a National Research Foundation of Korea Grant from the Korean Government (MSIT; Ministry of Science and ICT) (NRF-2021M1A5A1075524) (KOPRI-PN22014). This research was also supported by the Korea Polar Research Institute (KOPRI-PE22160).

Institutional Review Board Statement: Not applicable.

Informed Consent Statement: Not applicable.

Data Availability Statement: Not applicable.

Acknowledgments: We would like to thank the staff at BL-5C of the Pohang Accelerator Laboratory (Pohang, Korea) for their kind help with the X-ray diffraction data collection.

Conflicts of Interest: The authors declare no conflict of interest.

References

1. Marceau, A.H. Functions of single-strand DNA-binding proteins in DNA replication, recombination, and repair. In *Methods in Molecular Biology*; Walker, J.M., Ed.; Humana Press: Totowa, NJ, USA, 2012; Volume 922, pp. 1–21.
2. Lohman, T.M.; Ferrari, M.E. *Escherichia coli* single-stranded DNA-binding protein: Multiple DNA-binding modes and cooperativities. *Annu. Rev. Biochem.* **1994**, *63*, 527–570. [CrossRef]
3. Mishra, G.; Levy, Y. Molecular determinants of the interactions between proteins and ssDNA. *Proc. Natl. Acad. Sci. USA* **2015**, *112*, 5033–5038. [CrossRef] [PubMed]
4. Pestryakov, P.E.; Lavrik, O.I. Mechanisms of single-stranded DNA-binding protein functioning in cellular DNA metabolism. *Biochemistry (Moscow)* **2008**, *73*, 1388–1404. [CrossRef] [PubMed]
5. Theobald, D.L.; Mitton-Fry, R.M.; Wuttke, D.S. Nucleic acid recognition by OB-fold proteins. *Annu. Rev. Biophys. Biomol. Struct.* **2003**, *32*, 115–133. [CrossRef] [PubMed]
6. Flynn, R.L.; Zou, L. Oligonucleotide/oligosaccharide-binding fold proteins: A growing family of genome guardians. *Crit. Rev. Biochem. Mol. Biol.* **2010**, *45*, 266–275. [CrossRef] [PubMed]
7. Antony, E.; Lohman, T.M. Dynamics of *E. coli* single stranded DNA binding (SSB) protein-DNA complexes. *Semin. Cell Dev. Biol.* **2019**, *86*, 102–111. [CrossRef] [PubMed]
8. Shinn, M.K.; Kozlov, A.G.; Nguyen, B.; Bujalowski, W.M.; Lohman, T.M. Are the intrinsically disordered linkers involved in SSB binding to accessory proteins? *Nucleic Acids Res.* **2019**, *47*, 8581–8594. [CrossRef]
9. Costes, A.; Lecointe, F.; McGovern, S.; Quevillon-Cheruel, S.; Polard, P. The C-terminal domain of the bacterial SSB protein acts as a DNA maintenance hub at active chromosome replication forks. *PLoS Genet.* **2010**, *6*, e1001238. [CrossRef]
10. Bianco, P.R.; Pottinger, S.; Tan, H.Y.; Nguyenduc, T.; Rex, K.; Varshney, U. The IDL of *E. coli* SSB links ssDNA and protein binding by mediating protein–protein interactions. *Protein Sci.* **2017**, *26*, 227–241. [CrossRef]
11. Liu, T.; Huang, J. Replication protein A and more: Single-stranded DNA-binding proteins in eukaryotic cells. *Acta Biochim. Biophys. Sin.* **2016**, *48*, 665–670. [CrossRef]
12. Raghunathan, S.; Kozlov, A.G.; Lohman, T.M.; Waksman, G. Structure of the DNA binding domain of *E. coli* SSB bound to ssDNA. *Nat. Struct. Biol.* **2000**, *7*, 648–652. [CrossRef] [PubMed]
13. Fan, J.; Pavletich, N.P. Structure and conformational change of a replication protein A heterotrimer bound to ssDNA. *Genes Dev.* **2012**, *26*, 2337–2347. [CrossRef] [PubMed]
14. Ashton, N.W.; Bolderson, E.; Cubeddu, L.; O'Byrne, K.J.; Richard, D.J. Human single-stranded DNA binding proteins are essential for maintaining genomic stability. *BMC Mol. Biol.* **2013**, *14*, 9. [CrossRef] [PubMed]
15. Rapley, R. Enhancing PCR amplification and sequencing using DNA-binding proteins. *Mol. Biotechnol.* **1994**, *2*, 295–298. [CrossRef]
16. Dąbrowski, S.; Kur, J. Cloning, overexpression, and purification of the recombinant his-tagged SSB protein of *Escherichia coli* and use in polymerase chain reaction amplification. *Protein Expr. Purif.* **1999**, *16*, 96–102. [CrossRef]
17. Perales, C.; Cava, F.; Meijer, W.J.J.; Berenguer, J. Enhancement of DNA, cDNA synthesis and fidelity at high temperatures by a dimeric single-stranded DNA-binding protein. *Nucleic Acids Res.* **2003**, *31*, 6473–6480. [CrossRef]

18. Ducani, C.; Bernardinelli, G.; Högberg, B. Rolling circle replication requires single-stranded DNA binding protein to avoid termination and production of double-stranded DNA. *Nucleic Acids Res.* **2014**, *42*, 10596–10604. [CrossRef]
19. Cavicchioli, R.; Charlton, T.; Ertan, H.; Omar, S.M.; Siddiqui, K.S.; Williams, T.J. Biotechnological uses of enzymes from psychrophiles. *Microb. Biotechnol.* **2011**, *4*, 449–460. [CrossRef]
20. Lee, Y.M.; Shin, S.C.; Baek, K.; Hwang, C.Y.; Hong, S.G.; Chun, J.; Lee, H.K. Draft genome sequence of the psychrophilic bacterium *Lacinutrix jangbogonensis* PAMC 27137T. *Mar. Genom.* **2015**, *23*, 31–32. [CrossRef]
21. Thompson, J.D.; Gibson, T.J.; Higgins, D.G. Multiple sequence alignment using clustalW and clustalX. In *Current Protocol in Bioinformatics*; Baxevanis, A.D., Pearson, W.R., Stein, L.D., Storomo, G.D., Yates, J.R., III, Eds.; John Wiley & Sons: Hoboken, NJ, USA, 2002; pp. 2–3. [CrossRef]
22. Kumar, S.; Stecher, G.; Li, M.; Knyaz, C.; Tamura, K. MEGA X: Molecular evolutionary genetics analysis across computing platforms. *Mol. Biol. Evol.* **2018**, *35*, 1547–1549. [CrossRef]
23. Kabsch, W. XDS. *Acta Crystallogr. D Biol. Crystallogr.* **2010**, *66*, 125–132. [CrossRef] [PubMed]
24. Shereda, R.D.; Kozlov, A.G.; Lohman, T.M.; Cox, M.M.; Keck, J.L. SSB as an organizer/mobilizer of genome maintenance complexes. *Crit. Rev. Biochem. Mol. Biol.* **2008**, *43*, 289–318. [CrossRef] [PubMed]
25. Letunic, I.; Bork, P. Interactive Tree of Life (iTOL) v5: An online tool for phylogenetic tree display and annotation. *Nucleic Acids Res.* **2021**, *49*, W293–W296. [CrossRef] [PubMed]
26. Shamoo, Y.; Friedman, A.M.; Parsons, M.R.; Konigsberg, W.H.; Steitz, T.A. Crystal structure of a replication fork single-stranded DNA binding protein (T4 gp32) complexed to DNA. *Nature* **1995**, *376*, 362–366. [CrossRef] [PubMed]
27. Bernstein, D.A.; Eggington, J.M.; Killoran, M.P.; Misic, A.M.; Cox, M.M.; Keck, J.L. Crystal structure of the *Deinococcus radiodurans* single-stranded DNA-binding protein suggests a mechanism for coping with DNA damage. *Proc. Natl. Acad. Sci. USA* **2004**, *101*, 8575. [CrossRef]
28. Wold, M.S. Replication protein A: A heterotrimeric, single-stranded DNA-binding protein required for eukaryotic DNA metabolism. *Annu. Rev. Biochem.* **1997**, *66*, 61–92. [CrossRef]
29. Norais, C.A.; Chitteni-Pattu, S.; Wood, E.A.; Inman, R.B.; Cox, M.M. DdrB protein, an alternative *Deinococcus radiodurans* SSB induced by ionizing radiation. *J. Biol. Chem.* **2009**, *284*, 21402–21411. [CrossRef]
30. Huang, Y.H.; Lee, Y.L.; Huang, C.Y. Characterization of a single-stranded DNA binding protein from *Salmonella enterica* serovar Typhimurium LT2. *Protein J.* **2011**, *30*, 102–108. [CrossRef]
31. Huang, Y.-H.; Chen, I.C.; Huang, C.-Y. Characterization of an SSB–dT25 complex: Structural insights into the S-shaped ssDNA binding conformation. *RSC Adv.* **2019**, *9*, 40388–40396. [CrossRef]
32. Kozlov, A.G.; Cox, M.M.; Lohman, T.M. Regulation of single-stranded DNA binding by the C termini of *Escherichia coli* single-stranded DNA-binding (SSB) protein. *J. Biol. Chem.* **2010**, *285*, 17246–17252. [CrossRef]
33. Su, X.-C.; Wang, Y.; Yagi, H.; Shishmarev, D.; Mason, C.E.; Smith, P.J.; Vandevenne, M.; Dixon, N.E.; Otting, G. Bound or free: Interaction of the C-terminal domain of *Escherichia coli* single-stranded DNA-binding protein (SSB) with the tetrameric core of SSB. *Biochemistry* **2014**, *53*, 1925–1934. [CrossRef] [PubMed]
34. Vagin, A.; Teplyakov, A. Molecular replacement with MOLREP. *Acta Crystallogr. D Biol. Crystallogr.* **2010**, *66*, 22–25. [CrossRef] [PubMed]
35. Vagin, A.A.; Steiner, R.A.; Lebedev, A.A.; Potterton, L.; McNicholas, S.; Long, F.; Murshudov, G.N. REFMAC5 dictionary: Organization of prior chemical knowledge and guidelines for its use. *Acta Crystallogr. D Biol. Crystallogr.* **2004**, *60*, 2184–2195. [CrossRef] [PubMed]
36. Afonine, P.V.; Grosse-Kunstleve, R.W.; Echols, N.; Headd, J.J.; Moriarty, N.W.; Mustyakimov, M.; Terwilliger, T.C.; Urzhumtsev, A.; Zwart, P.H.; Adams, P.D. Towards automated crystallographic structure refinement with phenix. refine. *Acta Cryst. D Biol. Cryst.* **2012**, *68*, 352–367. [CrossRef]
37. Emsley, P.; Lohkamp, B.; Scott, W.G.; Cowtan, K. Features and development of Coot. *Acta Cryst. D Biol. Cryst.* **2010**, *66*, 486–501. [CrossRef]

 MDPI

Article

New Insight into the Effects of Various Parameters on the Crystallization of Ribulose-1,5-Bisphosphate Carboxylase/Oxygenase (RuBisCO) from *Alcaligenes eutrophus*

Hui-Woog Choe [1,*] and Yong Ju Kim [2,3,4,*]

1. Department of Chemistry, College of Natural Science, Jeonbuk National University, Baekje-daero 567, Jeonju 54896, Korea
2. Department of Lifestyle Medicine, College of Environmental and Bioresource Sciences, Jeonbuk National University, Gobong-ro 79, Iksan 54596, Korea
3. Department of Oriental Medicine Resources, College of Environmental and Bioresource Sciences, Jeonbuk National University, Gobong-ro 79, Iksan 54596, Korea
4. Advanced Institute of Environment and Bioscience, College of Environmental & Bioresources Sciences, Jeonbuk National University, Gobong-ro 79, Iksan 54596, Korea

* Correspondence: hwchoe@jbnu.ac.kr (H.-W.C.); nationface@jbnu.ac.kr (Y.J.K.)

Abstract: Crystallization remains a bottleneck for determining the three-dimensional X-ray structure of proteins. Many parameters influence the complexity of protein crystallization. Therefore, it is not easy to systematically examine all of these parameters individually during crystallization because of a limited quantity of purified protein. We studied several factors that influence crystallization including protein concentration, pH, temperature, age, volume of crystallization, inhibitors, metal ions, seeding, and precipitating agents on RuBisCO samples from *Alcaligenes eutrophus* which are not only freshly purified, but are also dissolved both individually and in combination from microcrystals and precipitated droplets of recycled RuBisCO. Single-, twin-, and/or microcrystals are dependent upon the concentration of RuBisCO by both RuBisCO samples. The morphology, either orthorhombic- or monoclinic-space group, depends upon pH. Furthermore, ammonium sulfate($(NH_4)_2SO_4$) concentration at 20 °C (22% saturated) and/or at 4 °C (28% saturated) affected the crystallization of RuBisCO differently from one another. Finally, the age of RuBisCO also affected more uniformity and forming sharp edge during crystallization. Unexpected surprising monoclinic RuBisCO crystals were grown from dissolved microcrystals and precipitated droplets recycled RuBisCO samples. This quaternary RuBisCO single crystal, which contained Mg^{2+} and HCO_3 for an activated ternary complex and is inhibited with a transition substrate analogue, CABP (2-carboxyarabinitol-1,5-bisphosphate)$^-$, diffracts better than 2.2 Å. It is different from Hansen S. et al. reported RuBisCO crystals which were grown ab initio in absence of Mg^{2+}, HCO_3^- and CABP, a structure which was determined at 2.7 Å resolution.

Keywords: RuBisCO; single-; twin-; and/or microcrystals; crystallization; precipitating agents; crystallization influencing parameter; morphology; an activated ternary complex; a quaternary complex

Citation: Choe, H.-W.; Kim, Y.J. New Insight into the Effects of Various Parameters on the Crystallization of Ribulose-1,5-Bisphosphate Carboxylase/Oxygenase (RuBisCO) from *Alcaligenes eutrophus*. *Crystals* **2022**, *12*, 196. https://doi.org/10.3390/cryst12020196

Academic Editors: Borislav Angelov, Kyeong Kyu Kim and T. Doohun Kim

Received: 24 December 2021
Accepted: 25 January 2022
Published: 28 January 2022

1. Introduction

Proteins are relatively long, fragile chain molecules composed of 20 different amino acids. Their biological activities are optimal within a narrow range of temperature and pH [1]. They function in the cell as structural elements, catalysts, transport, regulators of various processes, messengers, receptors for messengers, cell markers, and defense against cells that carry foreign antigens. Some proteins bind to DNA (deoxyribonucleic acid) or RNA to regulate recombination, whereas others participate in the replication, transcription, or translation of genetic information [2,3].

Probably the most important proteins are the enzymes that act as catalysts during cell metabolism. They recognize a specific molecule, the substrate, and bind to it in a dynamic equilibrium. The characteristics of an enzyme reside in the fact that it can chemically change a substrate by lowering activation energy under physiological conditions. Usually, this change accompanies the formation or cleavage of a covalent chemical bond. Thus, the substrate can be broken down into two or more parts: a chemical group can be attached, or binding patterns present in substrate molecules may be rearranged. The reaction mechanism of an enzyme is divided into three steps: the enzyme binds to its component, a chemical reaction takes place, and the altered substrate (product) disappears again. All three of these steps are reversible.

One of the main goals in protein research is to clarify the structure and reaction mechanisms of these molecules. The only method used to completely determine the structure of a protein in detail is X-ray crystallographic analysis [4,5]. Approximately 85% of the protein structures registered in the PDB have been determined by this method [6]. The aim of this study was to identify parameters that influence the crystallization of the enzyme, RuBisCO, derived from *A. eutrophus*. RuBisCO is the most abundant protein in nature and catalyzes the first reaction step of the Calvin-Benson-Bassham (CBB) cycle [7,8], which is involved in both of photosynthesis and photorespiration. Agricultural researchers are interested in increasing the production of crops by increasing photosynthesis, while decreasing photorespiration through genetic manipulation of the active sites of RuBisCO [9]. RuBisCO, in most bacteria as well as in higher plants, is composed of eight large (M.W. = 55,000) and eight small (M.W. = 14,000) subunits yielding a total molecular weight of 550,000 (L_8S_8). Both catalytic and active sites reside on the large subunit [10]. RuBisCO from *Rhodospirillum rubrum* consists of only two large subunits (L_2) [11]. However, it functions completely on its own in the cell without a small subunit. The role of the small subunit of RuBisCO, therefore, remains still unclear, although a mini review published about the role of the small subunit. It may regulate the structure or function of RuBisCO [12].

2. Material and Methods

Unless otherwise specified, materials from Sigma Aldrich (St. Louis, MO, USA) were used. Special chemicals or biochemicals have been purchased from the following companies, including ammonium peroxidosulfate (Merck, Taufkirchen, Germany), Amicon (MilliporeSigma, Burlington, MA, USA), Centricon 10 (MilliporeSigma, Burlington, MA, USA), RuBP(Ribulose 1,5-bisphosphate) (Sigma, Muenchen, Germany), NADH (Nicotinamide adenine dinucleotide), GDH (Glycerol-3-Phosphate Dehydrogenase)/TIM (triosephosphate isomerase), GAPDH (Glyceraldehyde 3-phosphate dehydrogenase)/PGK (Phosphoglycerate kinase), ATP (Adenosine triphosphate), EDTA (Ethylene-diamine-tetraacetic acid), PMSF (Phenylmethylsulfonyl fluoride), DTE (1,4-Dithioerythritol) (Roche, Basel, Switzerland).

Some of the experiments were undertaken at the Institute of Crystallography, Free University, Berlin, Germany and repeated at the laboratory of protein structure at the Department of Oriental Medicine Resources, College of Environmental and Bioresource Sciences, Jeonbuk National University, Republic of Korea.

2.1. Cell Cultivation and Purification of RuBisCO

A. eutrophus (ATCC 17699) was cultivated according to a culture manual and the harvested wet cells were collected by centrifugation (Hanil supra 30K, A50S-6, 4000 rpm, 4 °C, 15 min) and stored at −60 °C until use. For the isolation of RuBisCO, 10 g of wet cells were thawed and suspended in 100 mL isolation buffer (20 mM Tris/HCl, 50 mM $NaHCO_3$, 10 mM $MgCl_2$, 1 mM EDTA, 1 mM DTE, pH 8.0) with addition of 1/10 volume of 10 mM PMSF in ethanol and 10 mg DNase I. The cells were ruptured by sonication six times at 25 watts for 30 s with a 1-min break using an ultrasonicator R-4710-10 (Cole-Parmer, Vernon Hills, IL, USA) in an ice water bath. The samples were then centrifuged (Beckman coulter Optima XE-90, SW-28, 25,000 rpm, 4 °C, 30 min) and the pellets were discarded. Next, 500 µL were taken out from the supernatant and the absorption was measured at 260 and 280 nm using a UV-vis spectrometer (NEOGEN NEO-S490, Lansing, MI, USA). The O.D. (optical density) ratio (280/260) nm was approximately 0.84. According to the method of Warburg [13], the total amount of protein was approximately 2 g. Next, 80 mg protamine sulfate were added to eliminate nucleic acids and the samples were centrifuged (Hanil supra 30K, A50S-6, 20,000 rpm, 4 °C, 20 min). The resulting supernatant was mixed with 25% $(NH_4)_2SO_4$ (percentages for $(NH_4)_2SO_4$ in the description usually always refer to saturated = 100% solution). The samples were then re-centrifuged. The supernatant was adjusted to 40% saturation with $(NH_4)_2SO_4$ and the suspensions were kept at 4 °C for overnight. The next day, the samples were centrifuged (Hanil supra 30K, A50S-6, 20,000 rpm, 4 °C, 20 min). The supernatant was discarded and the pellets were solubilized with 10–20 mL of isolation buffer. Dialysis tubes (cut-off 6000–8000 Daltons) were used to desalt the samples. Before dialysis, the tubes were boiled for one hour in a solution containing 1 mM EDTA and 100 mM $NaHCO_3$. The samples were then dialyzed four times against the isolation buffer and the resulting solution served as a crude extract for RuBisCO enrichment.

The fresh purchased column materials were directly used for RuBisCO isolation, however, for recycling the column materials, the resin was washed on a glass filter with 1 M sodium acetate (pH 3, adjusted with acetic acid) until the pH of the eluate was 3. Thereafter, it was washed with 0.5 M NaOH to a pH of 14. It was re-washed with 1 M sodium acetate and then with isolation buffer until the pH value of the eluate was 8. The column material was finally degassed for 30 min in a suction flask. Before packing, the column material was stored at 4 °C.

The column material (DEAE-sepharose CL-6B) was poured into a column (d = 2 cm, h = 45 cm), which was placed vertically in a cold room (4 °C). A long glass rod was used so that no air bubbles could rise. Subsequently, the column was equilibrated with isolation buffer. The flow rate of the column was 1 mL/min. Then, 10–20 mL samples, which were dialyzed four times against the isolation buffers, were placed on the column and washed overnight with isolation buffer until no protein was detectable in the eluate. The proteins bound in the column material were then eluted by with a KCl gradient (0–0.4 M KCl) (Figure S1a). The eluate was divided into fractions of 7.5 mL each in the fractions in which enzyme activity was present. Purity was assessed by sodium dodecylsulfate polyacrylamide gel electrophoresis (SDS-PAGE) (Figure S1b).

2.2. Determination of RuBisCO Concentration

The protein concentration of the crude extract was determined by measuring the absorption at 260 and 280 nm using a UV-vis spectrometer (NEOGEN NEO-S490). The protein concentration of the purified RuBisCO was measured according to Bowien B. et al. [14] by absorbance at 280 nm (The absorption coefficient at 280 nm of RuBisCO solution was 1 mg/mL = 1.22).

2.3. Determination of RuBisCO Activity

To determine enzyme activity, the modified method of Racker E. [15] was used. For this assay the reagents shown in Table 1 were mixed. The solution was filled to 1000 µL with H_2O and the enzyme solution to be analyzed. The change in absorbance at 340 nm was recorded over time for 3 min.

2.4. Crystallization of RuBisCO

Hanging drop, sitting drop vapor diffusion, and microdialysis methods [16] were used for the crystallization of RuBisCO from *A. eutrophus*. For the salts in RuBisCO crystallization of this study, 100% saturated solutions in isolation buffer were used as stock solution. For polyethylenglycol (PEG) 6000, 40% solutions in buffer (w/v) were used as stock solutions. 2-methyl-2,4 pentandiol (MPD) solutions were prepared directly by using buffers. Among the precipitating agents tested, $(NH_4)_2SO_4$ was shown to be the most suitable; therefore, it was used as a precipitating agent for testing the effects of parameters influenced by the crystallization for this study.

Table 1. RuBisCO activity reagents and volume.

Reagent	Volume
1M Tris/HCl	50 μL
6 mM NADH	20 μL
0.1 M GSH	50 μL
0.5 M $KHCO_3$	150 μL
25 mM RUBP	20 μL
0.1 M ATP	120 μL
0.5 M $MgCl_2$	20 μL
10 mg/mL GDH/TIM	10 μL
6 mg/mL GAP-DH/PGK	20 μL
Total	460 μL

3. Results and Discussion

Protein crystallization is the process by which a metastable solid form, where highly purified homogeneous protein molecules are three dimensionally perfectly arranged into a protein crystal. In another word, it is a process of protein crystal formation via mechanisms of protein crystal growth. The protein crystallization point lies in general just below the precipitating point of a protein. There are many parameters that influence by the protein crystallization. We represent the results from the effects of parameters that influenced by RuBisCO crystallization from *A. eutrophus*. In the presence of Mg^{2+}, HCO_3^- activated ternary complex, and substrate analogue, the CABP inhibited the quaternary RuBisCO single crystal was formed. This resulted from CABP inhibited microcrystal droplets and precipitate droplets that had been dissolved in an isolation buffer over a period of days at 4 °C without mechanical intervention. The dissolved RuBisCO samples were then purifieded with a gel filtration to separate the denatured RuBisCO before crystallization. These were then compared with those from the freshly purified RuBisCO crystals.

3.1. *Effect of Temperature and pH Values*

The dependence of protein solubility on temperature resulted from the change in the acid/base reaction constant of the protein side chains as a function of temperature [17,18]. In addition, the pKa values of the ionizable groups were strictly related to the median ionic strength.

To correlate this phenomenon with salt and the change in precipitation characteristics, changes in conductivity with respect to concentration were determined and the effect of some parameters (temperature, pH, precipitant) on crystallization of RuBisCO from *A. eutrophus* was confirmed by electrical conductivity as shown in Figure 1. This method is relatively simple and different from the method described in a review by Nanev C.N. [19].

Fresh purified and/or recycled from microcrystals and precipitates droplets of RuBisCOs from *A. eutophus* were crystallized separately with 22% saturated $(NH_4)_2SO_4$ as a precipitating agent at 20 °C, while they were crystallized with 28% saturated $(NH_4)_2SO_4$ at 4 °C. The conductivity values for 22% saturated $(NH_4)_2SO_4$ at 20 °C and those of 28% $(NH_4)_2SO_4$ at 4 °C were very similar to one another between pH 7.0 and 8.0. This suggests that the conductivity values can provide information as to how one can obtain crystals with different salts as precipitating agents. Crystallization conditions of other salts such as Na_2SO_4 or $MgSO_4$ as precipitating agents shown in Table 2 were resulted from the conductivity measurement experiment.

The monoclinic morphology of RuBisCO crystals from *A. eutrophus* as decribed Section 3.2 in detail, is an unexpected result of the temperature effect on protein crystallization. The linbro plate which contained precipitated RuBisCO droplet on the concave of the bridge had been transferred from a temperature at 20 °C into a cooling room at 4 °C. This process might be induced the solubility change of RuBisCO sample in the droplet. The temperature affected the normal or retrograde solubility of RuBisCO samples. This might be a scientific plausible explanation how this monoclinic morphology of RuBisCO crystals could grow. The other plausible explanation in detail has been described in Section 3.2.

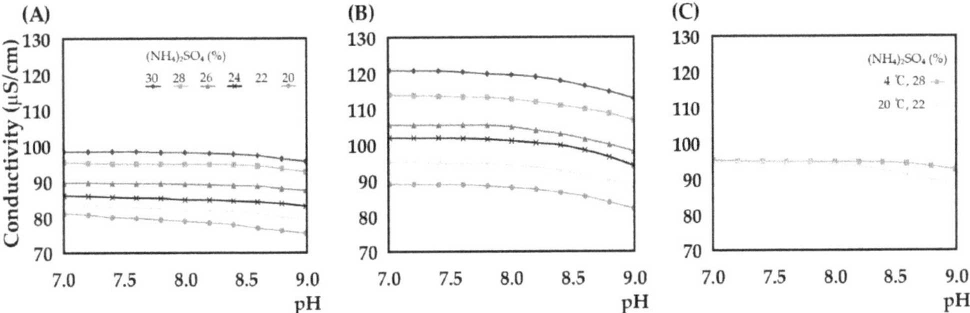

Figure 1. The electrical conductivity depends on pH, temperature and concentration of the precipitate. (**A**) The conductivity values (pH 7–9) depending on various $(NH_4)_2SO_4$ concentrations at 4 °C; (**B**) Those values at 20 °C (**C**) The overlapping conductivity values between 28% saturated $(NH_4)_2SO_4$ at 4 °C and those of 22% $(NH_4)_2SO_4$ at 20 °C. The conductivity values depended upon $(NH_4)_2SO_4$ concentrations were represented differently marked lines in the broken line graphs.

Table 2. Crystallization conditions and methods from both RuBisCO samples.

Precipitating Agents	Range of Precipitating Agents	Methods
$(NH_4)_2SO_4$	22–28% saturated in the isolation buffer	Sitting drop
Na_2SO_4	20–25% in isolation buffer	Microdialysis
$MgSO_4$	22–28% in isolation buffer	Sitting drop
MPD	25–30% in isolation buffer	Microdialysis
PEG 6000	10–15% in isolation buffer	Sitting drop

pH influences the nature of protein-protein interactions, which modify the potential for salt bridge and hydrogen bond formation. This is important for the formation of specific crystal contacts [20]. RuBisCO from *A. eutophus* showed a maximal enzyme activity in the range of pH 7.8–8.2 [21]. The morphology of orthorhombic and/or monoclinic RuBisCO crystals was dependent upon pH of crystallization conditions.

The first RuBisCO with Mg^{2+} and HCO_3^- activated ternary complex was crystallized at room temperature by Bowien B. et al. [22]. Between pH 7.0 and pH 8.4 the quaternary RuBisCO with Mg^{2+}, HCO_3^- and CABP was crystallized as orthorhombic which has been reported by Pal G.P. et al. [23]. The same quaternary RuBisCO samples either from fresh prepared or dissolved both from microcrystals droplets and precipitated droplets through a gel filtration recycled, were crystallized as monoclinic beyond pH 8.4 as represented in Figures 2, 3D and S2. This crystal morphology is new and indicates that pH change can induce to other morphologies of protein crystals. Whether this new crystal morphology has a merit for 3D structure determination or not is another matter. This result suggests that pH changes mainly influence to growing other morphologies of protein crystals.

After the crystallization setup, we observed the droplets on the concave of the bridge which were sealed with cover glasses, manually every two days for three weeks under a microscope. Results were recorded in tables in the laboratory notebook. Photos were occasionally taken and recorded. To avoid confusion when controls were carried out, a serial number and setup date were indicated on the cover and the bottom of every linbro plate.

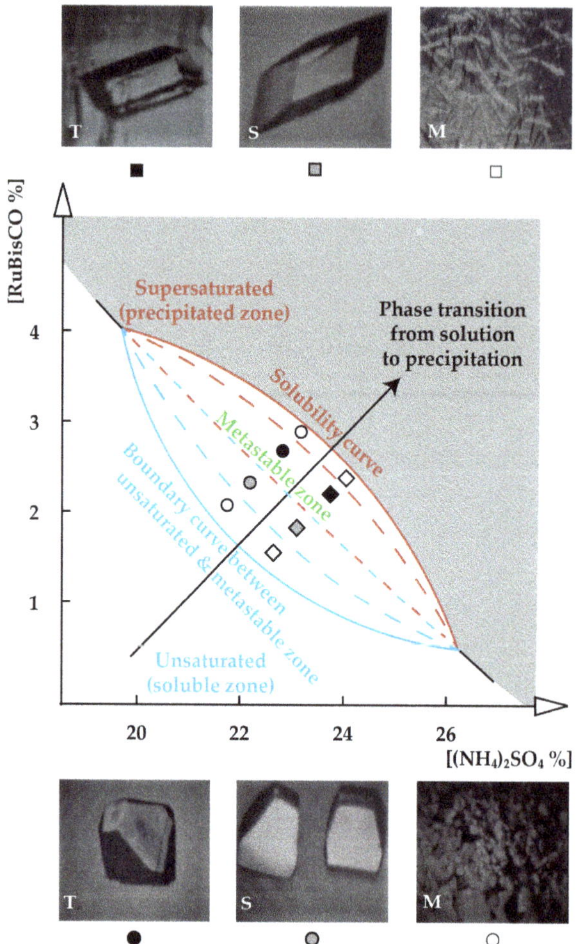

Figure 2. A canoe-shaped RuBisCO crystallization phase diagram The numbers of crystallization droplets which were observed orthorhombic twin crystals (T) are represented as ●, those of single crystals (S) are denoted as ◉, and those of microcrystals (M) represented as ○, while the numbers of crystallization droplets observed monoclinic twin crystals (T) are marked as ■, those of monoclinic single crystals (S) are represented as ▣, and those of microcrystals (M) are marked as □. The individual symbols in the canoe-shaped diagram denote the areas within the boundaries where characteristic morphologies of RuBisCO crystals were grown. Two upper –and down parts' areas, where microcrystals were grown in the diagram, are significantly different from each other. The extended lines from the two tips of the canoe-shaped diagram where do not exist the metastable zone, were observed either aggregates (precipitates) of RuBisCOs or occasionally salts crystals in the crystallization droplets.

After one week, the morphology differences could be observed after a control check setup. Two different morphologies of crystals were studied with X-ray to determine the space group. Regardless of the crystal morphologies, crystal growth was dependent on the number of nucleation seeds. Crystals moved within a droplet. The movement was either a zig-zag motion or they would sometimes roll. These movements are necessary in searching neighbor RuBisCO molecules for their growth. Such movement could be caused by a twin formation of RuBisCO crystals.

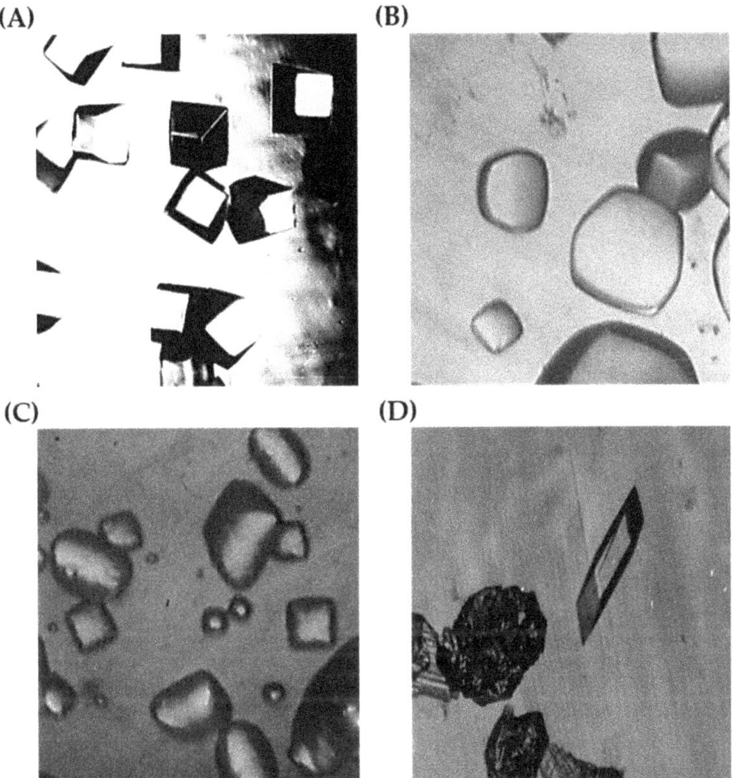

Figure 3. Differences in RuBisCO crystals depend upon the ages of enzymes. (**A**) Crystals from fresh purified RuBisCO, (**B**) Crystals from 3-months-old purified RuBisCO kept by 50% $(NH_4)_2SO_4$ suspension at 4 °C, (**C**) Crystals from 6 months old purified RuBisCO kept by 50% $(NH_4)_2SO_4$ suspension at 4 °C. (**D**) Monoclinic morphology of RuBisCO single crystal and poly-crystals which were recycled from microcrystals droplets and precipitated droplets.

3.2. Effect of Protein Concentration and Dependence on the Volume of the Crystallization Droplet

The effect of protein concentration was determined for a range of 1% to 4% at various pH values. After 3-weeks incubation, the formation of single-, twin- and/or micro crystals and even poly crystals depended heavily on protein concentration. It is interesting that the formations of single-, twin- and/or micro crystals and even poly crystals' morphology of orthorhombic and monoclinic of RuBisCO crystals were not significantly different at 20 °C from those at 4 °C. However, 22% saturated $(NH_4)_2SO_4$ concentrations at 20 °C and 28% saturated $(NH_4)_2SO_4$ concentration at 4 °C for the crystallization of RuBisCO from *A. eutrophus,* are fairly different from one another. As shown in Figure S2, the boundary between orthorhombic and monoclinic morphologies of RuBisCO crystals was pH 8.4. This boundary is marked as a vertical red dotted line. Figure S2 represents the diagram of the effects of pH, protein concentration complexity by the crystallization of RuBisCO from *A. eutrophus*. We represent a canoe-shaped RuBisCO crystallization phase diagram in Figure 2. This crystallization phase diagram was deduced from the raw data of Figure S2. This crystallization phase diagram is unique for RuBisCO from *A. eutrophus* and is not consistent with either the diagram of human carbonic anhydrase IX by Koruza K. et al. [24]

or nucleation and growth of protein crystals: general principles and assays in Methods in Enzymology [25]. We can understand easily from this canoe-shaped crystallization phase transition diagram both the salt-in effect and the salting out effect simultaneously which are required to crystallize a protein with precipitating agent as salts. A new insight into the well-established protein crystallization phase diagram might be arisen from the canoe-shaped RuBisCO crystallization phase diagram.

A surprising and unexpected result came by routine observation later in the month. From a 4% recycled RuBisCO sitting drop crystallization droplet was a perfectly grown single crystal and polycrystals as shown in Figure 3D. This unexpected RuBisCO crystal diffracts better than 2.2 Å. This crystal appeared from a droplet that was completely precipitated for a period of 3 weeks with regular checks. The linbro plate was left at 4 °C and was left unchecked for a month. It is hard to explain this phenomenon according to the theory that the crystallization point lies just below the precipitating point. A plausible explanation might be that we could not observe the nucleation of too tiny crystal nucleation seeds caused by too rapidly formed overclouded surrounded precipitates of RuBisCO molecules. The velocity of the transition from solution to precipitation was too rapid to observe and thereby grasp the crystallization point of RuBisCO as most of RuBisCO molecules themselves. However, after several days and even up to a month later, these nucleation seed crystals, in certain circumstances, came out from the surrounded precipitates and started to grow through sucking the precipitated, perfectly folded RuBisCO molecules. We could not observe this crystal growth under the microscope by through regular checks. During crystal growth, the surroundings of precipitates become clear because of decrease of the precipitates which are perfectly folded RuBisCO molecules. Because of the surrounding environment, growing RuBisCO crystal could not move freely. Therefore, rolling and zig-zag motion of RuBisCO crystals were not possible because of still much existing surrounded precipitates. Longer than a month, by chance could be seen wonderfully perfectly grown large single crystal as an outlier and too tiny crystal nucleation seeds which were located at almost the same position and suddenly covered with precipitates and by chance came out and started to grow finally as polycrystals of RuBisCO as represented in Figure 3D. Expectedly all precipitates surrounded crystal was totally disappeared. X-ray data analysis of this recycled unexpected RuBisCO crystal diffracted better than 2.2 Å is in progress.

Preliminary X-ray studies on the orthorhombic single and twin crystals from *A. eutrophus* have been reported [23]. However, monoclinic space group of RuBisCO crystals from *A. eutrophus* for the first resulted from this study. Preliminary X-ray data of differently crystallized RuBisCO crystals from *A. eutrophus* were represented in Table S1.

The optimal concentration varies with the protein used for the crystallization. Generally, the higher concentration of protein is more favorable as it appears to provide more opportunity for nucleation to occur. However, too high of a protein concentration can lead to an excess of nuclei and fewer large crystals and/or even polycrystals. Clearly, the more that is known about the solubility properties of the protein to be crystallized, the easier it will be to make the necessary adjustments for growing suitable single crystals [26] for X-ray analysis.

After determining the effect of protein concentration on crystallization, the amount of enzyme was varied (2 μL, 4 μL, 6 μL and 8 μL). The purpose of this experiment was to grow large single crystals. In crystallization sets where the volume of protein was 2–4 μL, a large number of small crystals grew. In contrast, with 8 μL, only 5–6 crystals grew in each crystallization set, which were usually 0.3 × 0.4 × 0.6 mm in size. The size of the crystals at a concentration of 2% was proportional to the amount of enzyme in the experiment.

3.3. *Effect of Precipitating Agents*

RuBisCO from *A. eutrophus* was crystallized using various precipitating agents and crystallization methods (Table 2).

Chemical compounds that reduce protein solubility are referred to as crystallizing (or precipitating) agents. They reinforce the attractions among bio-macromolecules and act either by altering the activity coefficient of water (salts) [27,28], changing the dielectric constant of the solvating medium (organic solvents), or increasing molecular crowding (high molecular weight polymers like PEG) [29]. Precipitants that act by different mechanisms show little exchangeability. Crystals obtained with one type of precipitant do not commonly form if the precipitant is changed with a functionally different one. However, it has been thoroughly demonstrated that combinations of mechanistically distinct precipitating agents can be synergistic and increase the probability of crystal growth.

3.4. Effect of Inhibitor and Dependence on RuBisCO Age

In general, it is more interesting to crystallize a protein together with a ligand as an apoenzyme. From the structure of such complexes, it may be possible to elucidate the biochemical reaction mechanism. For this reason, RuBisCO was co-crystallized with a transition state analog of its substrate. This analog CABP, binds very strongly to RuBisCO (K_d < 10 pM) [30]. Since CABP at a pH below 6 is present as a lactone, the ligand was added to an equal volume of 1 M Tris-HCl, pH 9, and incubated for 24 h at 20 °C. CABP was used in the crystallization experiments at a molar ratio of 1:16 (100% over excess), as each of the eight catalytic centers binds to CABP. When the buffered CABP solution was added to the protein-containing isolation buffer, the pH and ionic strength (conductivity) were changed. The 100% over excess of CABP was removed through a gel filtration, therefore in the outlier crystal in Figure 3D is the molar ratio between RuBisCO and CABP is 1:8. The K_d value of CABP to RuBisCO is extremely low, therefore dissociation of CABP from RuBisCO through a gel filtration is excluded. Holzenburg A. et al. presented 5 Å 3D structure model for the *A. eutrophus* RuBisCO in Nature (Table S2) [31]. They reported that the local 4-fold axes of the two L_4S_4 halves do not coincide but are shifted by 36 Å. This shift is caused by CABP bound to an activated ternary RuBisCO complex. However, there were many suspects about this structure model. Choe H.-W. et al. could not observe such a shift of 36 Å in solution between ternary activated RuBisCO and quaternary CABP bound inhibited RuBisCO through combining photon correlation and sedimentation analysis [32]. The structure of inactivated RuBisCO from *A. eutrophus* has been determined to 2.7 Å resolution by Hansen S. et al. and published [33]. They reported that the crystal structure of RuBisCO from *A. eutrophus* reveals a novel central eight-stranded beta-barrel formed by beta-strands from four subunits.

However, RuBisCO crystal by Hansen S. et al. and RuBisCO crystal in the current study are different from each other. RuBisCO crystals by Hansen S. et al. were crystallized in absence of Mg^{2+}, HCO_3^-, therefore totally ab initio inactivated RuBisCO (Table S2) [33]. The RuBisCO in current crystal has been crystallized in presence of Mg^{2+}, HCO_3^-, and inactivated through binding of CABP. This is a quaternary structure of RuBisCO. The current representing crystal structure analysis is required for clarification of a 36 Å shift between the two L_4S_4 halves through CABP bound to activated RuBisCO complex which could not observed by Choe H.-W. et al. in solution between ternary activated RuBisCO and quaternary CABP bound inhibited RuBisCO through combining photon correlation and sedimentation analysis. Presently, there are still not better than 5 Å resolution X-ray structures either with Mg^{2+}, HCO_3^- activated ternary complex or a Mg^{2+}, HCO_3^- and CABP inactivated quaternary complex of X-ray 3D structures from *A. eutrophus* [32].

RuBisCO from *A. eutrophus* lost less than 10% of its activity in a 50% $(NH_4)_2SO_4$ suspension in isolation buffer within 6 months at 4 °C. Crystallization experiments were conducted with protein stored for different lengths of time. It is clear that the quality of the crystals clearly depends on storage time (Figure 3A–C). Using fresh protein, crystals were obtained that showed more uniformity, better morphology, and were more suitable for X-ray study compared with crystals formed from aged proteins. The recycled CABP bound RuBisCO samples from dissolving microcrystals and precipitated droplets within a month could be grown and were not significantly different from the fresh purified RuBisCO samples. CABP bound RuBisCO samples which were dissolved from microcrystals and precipitated droplets, have been undertaken a gel filtration to separate the denatured

RuBisCO before. As an outlier recycled CABP bound RuBisCO crystal diffracts better than 2.2 Å. This crystal picture is represented in Figure 3D. This crystal picture represents how the uniformity of crystals can be improved by dissolving the aged microcrystals and precipitated samples through a simple gel filtration to separate the denatured proteins. This simple gel filtration might be a clue to grow a much better diffracted RuBisCO crystal. This sample was a homogeneous stoichiometric exact quaternary complex of RuBisCO. This sample was contained neither the denatured RuBisCOs nor the excess of CABP through a simple gel filtration.

3.5. Effect of Metal Ions

The active form of RuBisCO is a ternary complex with an allosteric effector, CO_2 [34], and a divalent metal ion [35–37]. As a Me^{2+} ion, Mg^{2+} exhibited the highest carboxylation activation. For the oxygenase reaction, however, Mn^{2+} and Co^{2+} as cofactors were more effective [38,39], which indicates different binding sites for the cations in the two reactions. For this reason, experiments were done using Mn^{2+} or Co^{2+} ions instead of Mg^{2+} for crystallization. Both ternary and quaternary complexes were crystallized with 10 mM $Co(NO_3)_2$ and $MnC1_2$, respectively, instead of $MgC1_2$ in the isolation buffer. With Mn^{2+} and Co^{2+}, RuBisCO crystals of the same morphology were formed as with Mg^{2+} (Figure 3A). The crystals of the complex with Mn^{2+} exhibited round edges. For the flame tests, Mn^{2+} or Co^{2+} contained RuBisCO crystals were washed thoroughly with isolation buffer without metal ions, respectively. The flame test was performed with dissolved crystal solutions. We could see the characteristic flame colors of Mn^{2+} or Co^{2+}. We, therefore, are sure RuBisCO crystal bound Mn^{2+} or Co^{2+} ions in crystals. However, showing the positions of each metal ion in the RuBisCO molecule is beyond the scope of this study, although it is very interesting question. The refined crystal 3D structures of both BuBisCO crystals either with Mn^{2+} or Co^{2+} will give the exact positions in the RuBisCO structures.

A procedure for the use of additives has recently been proposed [40,41]. In this technique, known as the cross-influence procedure, each crystallization trial utilizes four droplets containing equal volumes of the precipitating agent. The protein is added to one of the droplets, whereas additives (metallic salts) are placed in the others. Then, all drops are left to equilibrate against the same reservoir. In some cases, the ions are essential for biological activity and contribute to the maintenance of certain structural features of the protein. In other cases, metal ions stabilize intermolecular contacts in the crystal. Studies have shown that the application of biocompatible water-soluble ionic liquids, organic salts, and salts with melting points at or below 20 °C as crystallization additives provides very interesting results [42,43].

3.6. Seeding with Crystal Nucleus

It is often desirable to reproduce previously grown crystals of a protein in which either the formation of nuclei is limited or spontaneous nucleation occurs at such a profound level of supersaturation that poor growth patterns result. In such cases, it is desirable to induce growth in a directed fashion at low levels of supersaturation. This can sometimes be accomplished by seeding a metastable, supersaturated protein solution with crystals from earlier trials. These seeding techniques [44,45] fall into two categories: those employing microcrystals as seeds and those using larger macro seeds. For both methods, the fresh solution to be seeded should only be slightly supersaturated so that controlled, slow growth occurs. The two approaches have been described elsewhere in some detail [46,47].

The purpose of seeding is to limit the number of crystals in the crystallization set from the beginning. This increases the likelihood of obtaining large crystals. The collected microcrystals were washed with mother liquor and diluted (1:1000–1:5000) [48,49]. At 1 μL of the seeding, RuBisCO crystals grew, the quality of which was however not different from crystals obtained by other methods. It is recommended to seed at accurate time after set up the crystallization, 3–4 days after set up was the best time to seed by RuBisCO.

It depended on the protein concentration trials in the area of microcrystals without seeding, was more successful than those of high concentration for the RuBisCO crystallization. Caution is required when using crosslinked crystals or long kept crystals as they were not suitable for a seeding experiment. From freshly purified RuBisCO samples, grown microcrystals that are mechanically broken and diluted properly with crystallization buffer produced successfully grown single crystals. There can always be an exception in experimental research. Such an outlier can be a clue for a surprising result.

It is therefore important to continuing the persistent attempts to examine the effects of various parameters on the crystallization of a protein. Crystallization is a physical phenomenon. Observations of perfectly grown crystals under a polarizing microscope leave the impression that crystallization might be regarded not only as a science, but as a work of art.

Supplementary Materials: The following supporting information can be downloaded at: https://www.mdpi.com/article/10.3390/cryst12020196/s1, Figure S1: (a) Chromatogram of RuBisCO using by DEAE-Sepharose Cl -6B column. The diagonal line indi-cates the KCl concentration gradient. The peaks from I to V were loaded on the SDS-PAGE. Peak V indicates the RuBisCO fractions (Fraction number 62-71). (b) SDS-PAGE. Samples were taken out from the main fractions after the purification by DEAE- sepharose Cl-6B column. The lanes indicate as follows. (1) Marker, (2) Crude extract, (3) Ammonium sulfate fractionation (40%), (4) Peak I, (5) Peak II, (6) Peak III, (7) Peak IV, (8) Peak V, and (9) Side cuts of Peak V from the profile of the chromatogram shown as Figure S1a respectively; Figure S2: Influence of protein concentration on the crystallization of RuBisCO; Table S1: Preliminary X-ray data of differently crystallized RuBisCOs from *A. eutrophus*; Table S2: X-ray 3D structures of RuBisCOs from *A. eutrophus*.

Author Contributions: Conceptualization, Data curation, Formal analysis, Investigation, Methodology, Resources, Validation, Visualization, Writing—original draft, Writing—review & editing, H.-W.C. and Y.J.K.; funding acquisition, Y.J.K. All authors have read and agreed to the published version of the manuscript.

Funding: This research was supported by the Bio & Medical Technology Development Program and ICT (NRF-2017M3A9F6029733) and Basic Science Research Program of the National Research Foundation (NRF) funded by the Ministry of Science (NRF-2021R1I1A3060013).

Institutional Review Board Statement: Not applicable.

Informed Consent Statement: Not applicable.

Data Availability Statement: Not applicable.

Acknowledgments: H.-W. Choe thanks JBNU and DGIST for his chair professorship in the period of 2014–2021 at JBNU and in the period of 2017–2019 at DGIST. He is currently working as an emeritus professor at the Department of Chemistry, College of Natural Science, JBNU, Korea.

Conflicts of Interest: The authors declare no conflict of interest.

References

1. Schulz, G.E.; Schirmer, R.H. *Principles of Protein Structure*; Springer Science & Business Media: Berlin/Heidelberg, Germany, 2013.
2. Doolittle, R.F. Proteins. *Sci. Am.* **1985**, *253*, 88. [CrossRef] [PubMed]
3. Keene, J.D. Ribonucleoprotein infrastructure regulating the flow of genetic information between the genome and the proteome. *Proc. Natl. Acad. Sci. USA* **2001**, *98*, 7018–7024. [CrossRef] [PubMed]
4. Fersht, A. *Enzyme Structure and Mechanism*; W. H. Freeman and Company: New York, NY, USA, 1977.
5. Ilari, A.; Savino, C. Protein structure determination by X-ray crystallography. *Bioinformatics* **2008**, *452*, 63–87.
6. Worldwide Protein Data Bank Consortium. Protein Data Bank: The single global archive for 3D macromolecular structure data. *Nucleic Acids Res.* **2019**, *47*, D520–D528. [CrossRef]
7. Sainis, J.; Jawali, N. Channeling of the intermediates and catalytic facilitation to Rubisco in a multienzyme complex of Calvin cycle enzymes. *Indian J. Biochem. Biophys.* **1994**, *31*, 215–220. [PubMed]
8. Ducat, D.C.; Silver, P.A. Improving carbon fixation pathways. *Curr. Opin. Chem. Biol.* **2012**, *16*, 337–344. [CrossRef] [PubMed]
9. Ślesak, I.; Ślesak, H. The activity of RubisCO and energy demands for its biosynthesis. Comparative studies with CO_2-reductases. *J. Plant Physiol.* **2021**, *257*, 153337. [CrossRef] [PubMed]

10. Andersson, I.; Backlund, A. Structure and function of Rubisco. *Plant Physiol. Biochem.* **2008**, *46*, 275–291. [CrossRef]
11. Choe, H.-W.; Jakob, R.; Hahn, U.; Pal, G.P. Crystallization of the activated ternary complex of ribulose-1, 5-bisphosphate carboxylase-oxygenase isolated from Rhodospirillum rubrum and from an Escherichia coli clone. *J. Mol. Biol.* **1985**, *185*, 781–783. [CrossRef]
12. Spreitzer, R.J. Role of the small subunit in ribulose-1, 5-bisphosphate carboxylase/oxygenase. *Arch. Biochem. Biophys.* **2003**, *414*, 141–149. [CrossRef]
13. Warburg, O. Isolation and crystallization of enolase. *Biochem. Z.* **1942**, *310*, 384–421.
14. Bowien, B.; Mayer, F.; Codd, G.; Schlegel, H. Purification, some properties and quaternary structure of the D-ribulose 1, 5-diphosphate carboxylase of Alcaligenes eutrophus. *Arch. Microbiol.* **1976**, *110*, 157–166. [CrossRef] [PubMed]
15. Racker, E. [29a] Ribulose diphosphate carboxylase from spinach leaves: Ribulose diphosphate + CO_2 + H_2O → 2 3-P-Glycerate. In *Methods in Enzymology*; Elsevier: Amsterdam, The Netherlands, 1962; Volume 5, pp. 266–270.
16. McPherson, A.; Gavira, J.A. Introduction to protein crystallization. *Acta Crystallogr. Sect. F Struct. Biol. Commun.* **2014**, *70*, 2–20. [CrossRef] [PubMed]
17. Rosenberger, F.; Howard, S.; Sowers, J.; Nyce, T. Temperature dependence of protein solubility—Determination and application to crystallization in X-ray capillaries. *J. Cryst. Growth* **1993**, *129*, 1–12. [CrossRef]
18. Chernov, A.; Komatsu, H. Principles of crystal growth in protein crystallization. In *Science and Technology of Crystal Growth*; Springer: Berlin/Heidelberg, Germany, 1995; pp. 329–353.
19. Nanev, C.N. Recent insights into the crystallization process; protein crystal nucleation and growth peculiarities; Processes in the Presence of Electric Fields. *Crystals* **2017**, *7*, 310. [CrossRef]
20. Dumetz, A.C.; Chockla, A.M.; Kaler, E.W.; Lenhoff, A.M. Effects of pH on protein–protein interactions and implications for protein phase behavior. *Biochim. Biophys. Acta (BBA)-Proteins Proteom.* **2008**, *1784*, 600–610. [CrossRef]
21. Anderson, L.E.; Fuller, R. Photosynthesis in Rhodospirillum rubrum: IV. ISOLATION AND CHARACTERIZATION OF RIBULOSE 1, 5-DIPHOSPHATE CARBOXYLASE. *J. Biol. Chem.* **1969**, *244*, 3105–3109. [CrossRef]
22. Bowien, B.; Mayer, F.; Spiess, E.; Pähler, A.; Englisch, U.; Saenger, W. On the structure of crystalline ribulosebisphosphate carboxylase from Alcaligenes eutrophus. *Eur. J. Biochem.* **1980**, *106*, 405–410. [CrossRef]
23. Pal, G.P.; Jakob, R.; Hahn, U.; Bowien, B.; Saenger, W. Single and twinned crystals of ribulose-1, 5-bisphosphate carboxylase-oxygenase from Alcaligenes eutrophus. *J. Biol. Chem.* **1985**, *260*, 10768–10770. [CrossRef]
24. Koruza, K.; Lafumat, B.; Nyblom, M.; Knecht, W.; Fisher, Z. From initial hit to crystal optimization with microseeding of human carbonic anhydrase IX—A case study for neutron protein crystallography. *Crystals* **2018**, *8*, 434. [CrossRef]
25. Feher, G.; Kam, Z. [4] Nucleation and growth of protein crystals: General principles and assays. *Methods Enzymol.* **1985**, *114*, 77–112. [PubMed]
26. Littlechild, J. Protein crystallization: Magical or logical: Can we establish some general rules? *J. Phys. D Appl. Phys.* **1991**, *24*, 111. [CrossRef]
27. Hofmeister, F. Zur lehre von der wirkung der salze. *Arch. Exp. Pathol. Pharmakol.* **1888**, *24*, 247–260. [CrossRef]
28. Kunz, W.; Henle, J.; Ninham, B.W. 'Zur Lehre von der Wirkung der Salze' (about the science of the effect of salts): Franz Hofmeister's historical papers. *Curr. Opin. Colloid Interface Sci.* **2004**, *9*, 19–37. [CrossRef]
29. McPherson, A., Jr. Crystallization of proteins from polyethylene glycol. *J. Biol. Chem.* **1976**, *251*, 6300–6303. [CrossRef]
30. Pierce, J.; Tolbert, N.; Barker, R. Interaction of ribulosebisphosphate carboxylase/oxygenase with transition-state analogs. *Biochemistry* **1980**, *19*, 934–942. [CrossRef] [PubMed]
31. Holzenburg, A.; Mayer, F.; Harauz, G.; Van Heel, M.; Tokuoka, R.; Ishida, T.; Harata, K.; Pal, G.; Saenger, W. Structure of D-ribulose-l, 5-bisphosphate carboxylase/oxygenase from Alcaligenes eutrophyus H16. *Nature* **1987**, *325*, 730–732. [CrossRef]
32. Choe, H.-W.; Georgalis, Y.; Saenger, W. Comparative studies of ribulose-1, 5-biphosphate carboxylase/oxygenase from Alcaligenes eutrophus H16 cells, in the active and CABP-inhibited forms. *J. Mol. Biol.* **1989**, *207*, 621–623. [CrossRef]
33. Hansen, S.; Vollan, V.B.; Hough, E.; Andersen, K. The crystal structure of Rubisco from Alcaligenes eutrophus reveals a novel central eight-stranded β-barrel formed by β-strands from four subunits. *J. Mol. Biol.* **1999**, *288*, 609–621. [CrossRef]
34. Badger, M.; Andrews, T. Effects of CO_2, O_2 and temperature on a high-affinity form of ribulose diphosphate carboxylase-oxygenase from spinach. *Biochem. Biophys. Res. Commun.* **1974**, *60*, 204–210. [CrossRef]
35. Lorimer, G.H.; Badger, M.R.; Andrews, T.J. The activation of ribulose-1,5-bisphosphate carboxylase by carbon dioxide and magnesium ions. Equilibria, kinetics, a suggested mechanism, and physiological implications. *Biochemistry* **1976**, *15*, 529–536. [CrossRef] [PubMed]
36. Paech, C.; Tolbert, N. Active site studies of ribulose-1,5-bisphosphate carboxylase/oxygenase with pyridoxal 5′-phosphate. *J. Biol. Chem.* **1978**, *253*, 7864–7873. [CrossRef]
37. Laing, W.A.; Christeller, J.T. A model for the kinetics of activation and catalysis of ribulose 1,5-bisphosphate carboxylase. *Biochem. J.* **1976**, *159*, 563–570. [CrossRef] [PubMed]
38. Robison, P.D.; Martin, M.N.; Tabita, F.R. Differential effects of metal ions on Rhodospirillum rubrum ribulosebisphosphate carboxylase/oxygenase and stoichiometric incorporation of bicarbonate (1-) ion into a cobalt (III)-enzyme complex. *Biochemistry* **1979**, *18*, 4453–4458. [CrossRef]
39. Wildner, G.F.; Henkel, J. Differential reactivation of ribulose 1,5-bisphosphate oxygenase with low carboxylase activity by Mn^{2+}. *FEBS Lett.* **1978**, *91*, 99–103. [CrossRef]

40. Tomčová, I.; Branca, R.M.M.; Bodó, G.; Bagyinka, C.; Kutá Smatanová, I. Cross-crystallization method used for the crystallization and preliminary diffraction analysis of a novel di-haem cytochrome c4. *Acta Crystallogr. Sect. F Struct. Biol. Cryst. Commun.* **2006**, *62*, 820–824. [CrossRef]
41. Tomčová, I.; Smatanová, I.K. Copper co-crystallization and divalent metal salts cross-influence effect: A new optimization tool improving crystal morphology and diffraction quality. *J. Cryst. Growth* **2007**, *306*, 383–389. [CrossRef]
42. Hekmat, D.; Hebel, D.; Joswig, S.; Schmidt, M.; Weuster-Botz, D. Advanced protein crystallization using water-soluble ionic liquids as crystallization additives. *Biotechnol. Lett.* **2007**, *29*, 1703–1711. [CrossRef]
43. Judge, R.A.; Takahashi, S.; Longenecker, K.L.; Fry, E.H.; Abad-Zapatero, C.; Chiu, M.L. The effect of ionic liquids on protein crystallization and X-ray diffraction resolution. *Cryst. Growth Des.* **2009**, *9*, 3463–3469. [CrossRef]
44. Bergfors, T. Seeds to crystals. *J. Struct. Biol.* **2003**, *142*, 66–76. [CrossRef]
45. Stura, E.A.; Wilson, I.A. Applications of the streak seeding technique in protein crystallization. *J. Cryst. Growth* **1991**, *110*, 270–282. [CrossRef]
46. Fitzgerald, P.M.; Madsen, N.B. Improvement of limit of diffraction and useful X-ray lifetime of crystals of glycogen debranching enzyme. *J. Cryst. Growth* **1986**, *76*, 600–606. [CrossRef]
47. Thaller, C.; Eichele, G.; Weaver, L.; Wilson, E.; Karlsson, R.; Jansonius, J. [9] Seed enlargement and repeated seeding. *Methods Enzymol.* **1985**, *114*, 132–135. [PubMed]
48. Thaller, C.; Weaver, L.; Eichele, G.; Wilson, E.; Karlsson, R.; Jansonius, J. Repeated seeding technique for growing large single crystals of proteins. *J. Mol. Biol.* **1981**, *147*, 465–469. [CrossRef]
49. Smit, J.D.G.; Winterhalter, K.H. Crystallographic data for haemoglobin from the lanceolate fluke Dicrocoelium dendriticum. *J. Mol. Biol.* **1981**, *146*, 641–647. [CrossRef]

MDPI

Communication

Secondary Structure and X-ray Crystallographic Analysis of the Glideosome-Associated Connector (GAC) from *Toxoplasma gondii*

Amit Kumar [1], Xu Zhang [1], Oscar Vadas [2], Fisentzos A. Stylianou [1], Nicolas Dos Santos Pacheco [2], Sarah L. Rouse [1], Marc L. Morgan [1], Dominique Soldati-Favre [2] and Steve Matthews [1,*]

1 Department of Life Sciences, Imperial College London, South Kensington, London SW7 2AZ, UK; amit.kumar@imperial.ac.uk (A.K.); x.zhang20@imperial.ac.uk (X.Z.); fisentzos.stylianou@outlook.com (F.A.S.); s.rouse@imperial.ac.uk (S.L.R.); rhodri.morgan@imperial.ac.uk (M.L.M.)

2 Department of Microbiology & Molecular Medicine, University of Geneva, 1 Rue Michel-Servet, 1211 Geneva, Switzerland; oscar.vadas@unige.ch (O.V.); nicolas.dossantospacheco@unige.ch (N.D.S.P.); Dominique.Soldati-Favre@unige.ch (D.S.-F.)

* Correspondence: s.j.matthews@imperial.ac.uk; Tel.: +207-5945-315

Abstract: A model for parasitic motility has been proposed in which parasite filamentous actin (F-actin) is attached to surface adhesins by a large component of the glideosome, known as the glideosome-associated connector protein (GAC). This large 286 kDa protein interacts at the cytoplasmic face of the plasma membrane with the phosphatidic acid-enriched inner leaflet and cytosolic tails of surface adhesins to connect them to the parasite actomyosin system. GAC is observed initially to the conoid at the apical pole and re-localised with the glideosome to the basal pole in gliding parasite. GAC presumably functions in force transmission to surface adhesins in the plasma membrane and not in force generation. Proper connection between F-actin and the adhesins is as important for motility and invasion as motor operation itself. This notion highlights the need for new structural information on GAC interactions, which has eluded the field since its discovery. We have obtained crystals that diffracted to 2.6–2.9 Å for full-length GAC from *Toxoplasma gondii* in native and selenomethionine-labelled forms. These crystals belong to space group $P2_12_12_1$; cell dimensions are roughly a = 119 Å, b = 123 Å, c = 221 Å, $\alpha = 90°$, $\beta = 90°$ and $\gamma = 90°$ with 1 molecule per asymmetric unit, suggesting a more compact conformation than previously proposed

Keywords: apicomplexa; *Toxoplasma gondii*; motility; invasion; glideosome-associated connector protein; X-ray crystallography

Citation: Kumar, A.; Zhang, X.; Vadas, O.; Stylianou, F.A.; Dos Santos Pacheco, N.; Rouse, S.L.; Morgan, M.L.; Soldati-Favre, D.; Matthews, S. Secondary Structure and X-ray Crystallographic Analysis of the Glideosome-Associated Connector (GAC) from *Toxoplasma gondii*. *Crystals* **2022**, *12*, 110. https://doi.org/10.3390/cryst12010110

Academic Editors: Kyeong Kyu Kim and T. Doohun Kim

Received: 27 December 2021
Accepted: 12 January 2022
Published: 15 January 2022

1. Introduction

There are more than 5000 species of apicomplexan parasites and many are etiological agents of major diseases that are a threat to global human and animal health, particularly in low-resource settings [1]. Most significant are malaria (*Plasmodium*), cryptosporidiosis (*Cryptosporidium*) and toxoplasmosis (*Toxoplasma*). The lifestyle of these obligate intracellular parasites involves crucial steps that depend on gliding motility such as host cell invasion, egress from infected cell and crossing of biological barriers [2].

These processes are dependent upon the orchestrated release of proteins from apical secretory organelles: micronemes and rhoptries. Following initial apical organelle secretion, a moving junction is formed that participates in the active penetration of host cells. In addition to invasion, parasites also use gliding motility to actively exit infected host cells during egress or migrate across biological surfaces, and in all cases, motility appears to be powered by the glideosome [3–5].

Current understanding broadly agrees upon a molecular architecture for the glideosome and explains how an actomyosin-motor drives motility in apicomplexan parasites. In this model, the TRAP/MIC family of adhesins targets ligands on the surface of the host cell to mediate apical attachment [6]. At the cytoplasmic side of the plasma membrane these adhesins are connected to the parasite actin filament network, while myosin motors drag adhesins through the plane of the plasma membrane towards the parasite posterior, and consequently pull the host cell membrane around the parasite. Myosin A (MyoA), a small divergent class XIV myosin, together with and glideosome-associated proteins (GAPs), act as the motor powering gliding motility [7]. MyoA is situated between the inner membrane complex (IMC) and parasite plasma membrane, and, together with the GAPs, it spans the two membranous structures.

The molecular component that bridges the adhesin to F-actin is a large novel protein termed the glideosome associated connector protein (GAC), which translocates with the moving junction from the parasite apex to the basal pole during gliding motility [8]. GAC is highly conserved across the entire Apicomplexa phylum and forms complexes with three binding partners (Figure 1A). The C-terminal region of GAC interacts with phosphatidic acid (PA) enriched membranes and its deletion results in a defective lytic-cycle phenotype. It has been suggested that GAC localisation is dependent upon PA generated by the lipid signalling cascade that regulates the apicomplexan Pleckstrin-homology domain protein (APH) to control microneme secretion [9,10]. PA signalling may ensure GAC is appropriately recruited only when a productive interaction can occur with adhesins during gliding motility. GAC also binds to microneme adhesin C-terminal tails at the plasma membrane's cytoplasmic face and therefore serves as the link to the surface adhesins. Thirdly, a direct connection to the glideosome is made via an interaction between GAC and the parasite actin filaments (F-actin). It has also been shown that proper connection of actin to the adhesins (via GAC) is more important for efficient motility and invasion than motor operation itself [11].

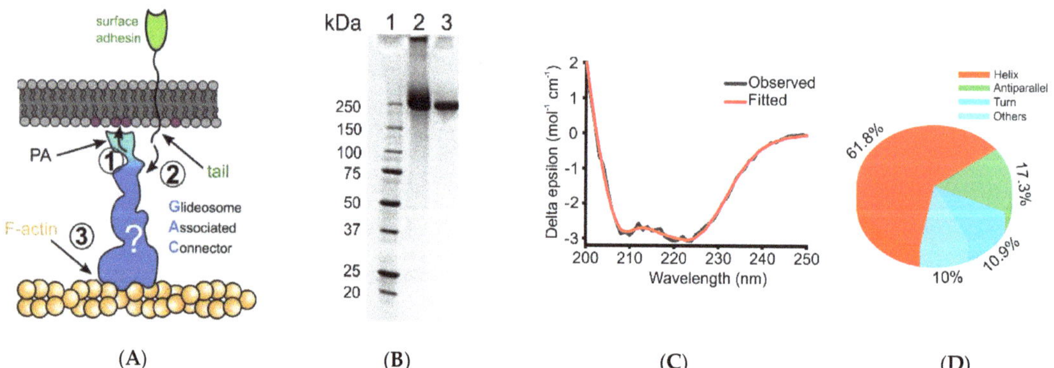

Figure 1. TgGAC interactions, purification and characterisation. (**A**) Schematic representation of GAC interactions with (1) the plasma membrane, (2) surface adhesins and (3) F-actin (**B**) SDS–PAGE of purified full-length TgGAC used for the crystallization trials. Lane 1: molecular-weight markers (kDa), lane 2: Ni-NTA eluted fraction and lane 3: eluted fraction from a Sephecryl S-300 HR SEC column. (**C**) CD spectrum of purified TgGAC measured at 298 K. Black spectrum showed the observed spectrum in 25 mM Tris.HCl, pH 8.0. Red spectrum indicated the fitted line for secondary structure analysis. (**D**) Secondary structure analysis based upon the CD spectrum as indicated in (**C**). Orange indicated the percentage helical content, green is beta-sheet, turn as blue and cyan as other contents in the GAC.

We set out to solve the structure of GAC from *T. gondii*. Initial efforts to generate high-resolution diffracting crystals of TgGAC proved challenging. Eventually, after optimisation of the purification conditions together with secondary structure analyses and extensive crystallization screening, we generated high quality crystals that diffract to 2.67 Å resolution.

This represents a key breakthrough in providing the first structural insight into the GAC architecture and function within the glideosome.

2. Materials and Methods

2.1. Protein Expression and Purification

Full-length TgGAC genes with TEV cleavable N-terminal 6xHis-tag has been cloned into the pET28a vector as previously described [8]. For selenomethionine labeling, the methionine auxotrophic *E. coli* strain B834(DE3) was transformed with the above mentioned vector. Cells were grown in the standard M9 minimal media containing the 50 μg/mL of methionine until the OD_{600nm}~1. The cells were harvested and resuspended in fresh M9 media without methionine, and incubated for 4 h at 37 °C. Seleno-L-Methionine (50 μg/mL) was subsequently added and incubated for a further 30 min. The GAC expression was started with 1 mM IPTG and further incubated overnight at 22 °C.

Cells were harvested and resuspended in 50 mM Tris (pH 8), 300 mM NaCl, 10 mM Imidazole and 5 mM TCEP, followed by lysis by sonication and centrifugation at 18,000 rpm for 60 min (Ti45 rotor, Beckman, Brea, CA, USA). TgGAC was then purified by nickel chromatography followed by gel filtration using a Sephecryl S-300 HR column (GE Healthcare, Chicago, IL, USA).

2.2. TgGAC Crystallization

Conditions for crystallization were initially screened by the sitting drop method of vapour diffusion at 20 °C and 4 °C using sparse matrix crystallization kits (Hampton research and Molecular Dimensions). MRC 96-well optimization plates (Molecular Dimensions) were utilised. Each drop was set with 100 nL protein solution and 100 nL reservoir solution using a Mosquito nanolitre high-throughput robot (TTP Labtech, Melbourn, UK). Reproducible protein crystals were obtained in 100 mM magnesium acetate, 100 mM sodium acetate, 6% PEG8000 and pH 5.0. These were manually optimised by screening over sodium acetate pH ranges of 4.0 to 5.0 in one dimension and a PEG8000 concentration gradient of 4%–10% in the second dimension. Crystallization was set-up at a concentration of 5–60 mg mL^{-1}.

2.3. Circular Dichroism (CD) Spectroscopy

CD spectra of GAC was recorded at a concentration of 0.8 mg mL^{-1} in a variety of solution conditions, in which buffer, pH and temperature were changed. Spectra were recorded in the wavelength range of 200–260 nm, with a scan length of 2 s per point. Four repeats were collected and averaged. Optimal spectra were recorded for TgGAC at 298 K in 25 mM Tris.HCl, pH 8.0. Data were collected and processed by Chirascan CD Spectrometer (Applied Photophysics Limited, Leatherhead, UK).

2.4. X-ray Data Collection and Processing

Crystals were mounted in a MicroLoop (MiTeGen), cryoprotected with 30% ethylene glycol for 5 s and immediately flash frozen in liquid nitrogen. Diffraction data from native crystals were collected on beamline I04 of the Diamond Light Source (DLS), UK. Data were processed with CCP4, dials [12–15] and scaled using dials.scale [16] within the Xia2 package [17]. Multiple-wavelength anomalous diffraction (MAD) data from a single SeMet labelled crystal were collected on beamline I04 of the Diamond Light Source at the following wavelengths: peak = 0.9795 Å, inflection = 0.9796 Å and remote = 0.9722 Å. Data were processed initially by AutoProc [18]. Substructure definition and initial model building were performed using AutoSHARP [19]. This was followed by manual building in Coot [20] and further refinement using Phenix Refine [21].

Data collection statistics are shown in Table 1. The content of the unit cell was analysed using the Matthews coefficient [22]. Molecular replacement (MR) attempts were carried out using computationally derived structures using the following servers: RaptorX, Alphafold and iTasser [23–25].

Table 1. X-ray diffraction data collection statistics.

Crystal	Native	SeMet
Space group	$P2_12_12_1$	$P2_12_12_1$
Cell dimensions (Å)	a = 120.63, b = 123.96, *c* = 221.86	a = 119.08, b = 123.60, c = 221.51
Angles (°)	α = 90.00, β = 90.00, γ = 90.00	α = 90.00, β = 90.00, γ = 90.00
Resolution (Å)	82.66–2.92 (2.97–2.92) *	110.75–2.67 (2.67–2.72) *
Wavelength (Å)	0.97950	0.97950
Total reflections	2,899,820 (140,928)	904,519 (44214)
Unique observations	72,969 (3564)	92,922 (4581)
Completeness (%)	100 (99–100)	100 (99.7–100)
Multiplicity	39.7 (39.5)	9.7 (9.7)
R_{pim} †	0.045 (1.486)	0.039 (0.861)
$<I>/\sigma I$	13.1 (88–0.5)	14 (47.2–0.8)
Molecules per asymmetric unit ‡	1	1
Solvent content (%)	56.8	56.8

* Values in parentheses correspond to the highest resolution shell. † $R_{merge} = \Sigma_{hkl} \Sigma_i | I_i(hkl) - <I(hkl)> | / \Sigma_{hkl} \Sigma_i I_i(hkl)$ where $<I(hkl)>$ is the mean intensity of the observations $I_i(hkl)$ of reflection *hkl*. ‡ Most probable value.

3. Results and Discussion

While the identity and function of crucial genetic components of the *T. gondii* life cycle and infectivity are known, a detailed mechanistic understanding of parasite motility and invasion remains limited. Despite GAC's essential role in efficient motility and invasion, no high-resolution experimental structural information is available. A small-angle X-ray scattering (SAXS) study presented TgGAC as a ~27 nm club-shaped molecule that stretches across the space between the parasitic plasma membrane and F-actin [8] (Figure 1A). However, this model is inconsistent with the latest understanding of the glideosome, as GAC would be unable to fit lengthways across this space together with the other essential components. To fully understand how GAC carries out its role, new experimental structural insight is required. We therefore isolated and purified the full-length TgGAC (Figure 1B). TgGAC contains 75 cysteines residues that are predicted not to participate in disulphide bonds. After assessment of structure and stability of GAC with CD spectroscopy (Figure 1C,D), we found that maintaining strict reducing conditions throughout purification was a crucial step in maintaining GACs full secondary structure (61.8% α-helix, 17.3% β-sheet, 10.8% turn and 10% others). Under these experimental conditions, crystals reliably grew to 100–200 μm^3 in size over the course of 5–7 days (Figure 2A). For the highest resolution, it was essential to acquire X-ray diffraction images immediately after harvesting and freezing. Frozen crystals stored for any longer than a few days showed a progressive deterioration in resolution. Crystals could be stored at room temperature in crystallisation wells for many weeks without loss in the number of high-resolution crystals, if diffraction data were acquired directly after harvesting.

Native diffraction data were collected to 2.92 Å (Figure 2B) and indexed in space group $P2_12_12_1$. Analysis of the crystal content indicated cell dimensions are a = 119 Å, b = 123 Å, c = 221 Å, α = 90°, β = 90° and γ = 90° with 1 molecule per asymmetric unit with a solvent content of 56%. This suggests that GAC adopts a more compact conformation than the ~27 nm (270 Å) long, club-shape measured from SAXS analysis, which would be better matched to the confined space between the plasma membrane and F-actin. Data-collection and processing statistics are listed in Table 1.

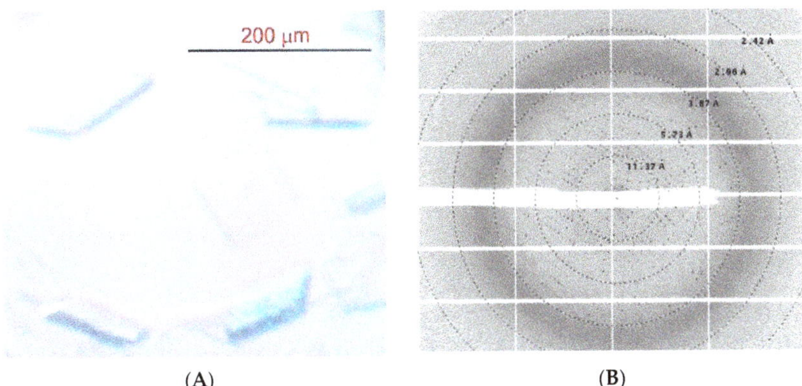

(A) (B)

Figure 2. TgGAC full length crystallography. (**A**) Representative native crystals of TgGAC. Scale bar represents 200 μm. (**B**) Diffraction image from a TgGAC crystal. Resolution rings are annotated.

Molecular replacement attempts were made using structures based on various prediction algorithms as search models: RaptorX, Alphafold and iTasser. No solutions were found. We subsequently prepared selenomethionine-substituted and heavy-atom derivatives to provide phases using anomalous dispersion techniques. A single crystal of selenomethionine-labelled TgGAC was successfully obtained in the same conditions as native protein. Incorporation of heavy atoms was monitored by fluorescence scan screening of crystals on the beamline prior to data collection. Successful incorporation was validated by detection of a distinct peak corresponding to the X-ray absorption edge for Selenium (0.9795 A or 12.6578 keV). The crystals of native and SeMet-derivatized TgGAC protein diffracted to 2.92 Å and 2.67 Å resolution, respectively. Matthews coefficient (VM = 2.70 $Å^3$ Da^{-1}) and solvent-content (VS = 56.8%) calculations indicated that one molecule was present in the asymmetric unit. The multiple-wavelength anomalous dispersion method was used to determine the initial phases of the SeMet-substituted TgGAC protein. The autoSHARP processing pipeline obtained 73 Se sites. The structure solution is currently under way and will be reported elsewhere with full statistics.

Author Contributions: Conceptualization, S.M. and D.S.-F.; methodology, A.K., O.V., S.L.R. and M.L.M.; formal analysis, A.K. and M.L.M.; investigation, A.K., X.Z., F.A.S., N.D.S.P.; writing—original draft preparation, S.M.; writing—review and editing, A.K., X.Z., F.A.S., S.L.R. and D.S.-F.; supervision, S.M. and D.S.-F.; project administration, S.M. and D.S.-F.; funding acquisition, S.M. and D.S.-F. All authors have read and agreed to the published version of the manuscript.

Funding: Leverhulme Trust awards to SJM (RPG_2018_107); SNSF funding to DSF (310030_185325).

Acknowledgments: The crystallization facility at Imperial College London which is supported by the Biotechnology and Biological Sciences Research Council (BB/D524840/1) and Wellcome Trust (202926/Z/16/Z). We are grateful to staff at Diamond Light Source beamline I04 for their help with data collection.

Conflicts of Interest: The authors declare no conflict of interest.

References

1. Meissner, M. The asexual cycle of apicomplexan parasites: New findings that raise new questions. *Curr. Opin. Microbiol.* **2013**, *16*, 421–423. [CrossRef]
2. Heintzelman, M.B. Gliding motility in apicomplexan parasites. *Semin. Cell Dev. Biol.* **2015**, *46*, 135–142. [CrossRef]
3. Frenal, K.; Dubremetz, J.F.; Lebrun, M.; Soldati-Favre, D. Gliding motility powers invasion and egress in Apicomplexa. *Nat. Rev. Microbiol.* **2017**, *15*, 645–660. [CrossRef]
4. Boucher, L.E.; Bosch, J. The apicomplexan glideosome and adhesins—Structures and function. *J. Struct. Biol.* **2015**, *190*, 93–114. [CrossRef] [PubMed]

5. Cowman, A.F.; Tonkin, C.J.; Tham, W.H.; Duraisingh, M.T. The Molecular Basis of Erythrocyte Invasion by Malaria Parasites. *Cell Host Microbe* **2017**, *22*, 232–245. [CrossRef] [PubMed]
6. Carruthers, V.B.; Tomley, F.M. Microneme proteins in apicomplexans. *Subcell. Biochem.* **2008**, *47*, 33–45. [CrossRef] [PubMed]
7. Powell, C.J.; Ramaswamy, R.; Kelsen, A.; Hamelin, D.J.; Warshaw, D.M.; Bosch, J.; Burke, J.E.; Ward, G.E.; Boulanger, M.J. Structural and mechanistic insights into the function of the unconventional class XIV myosin MyoA from Toxoplasma gondii. *Proc. Natl. Acad. Sci. USA* **2018**, *115*, E10548–E10555. [CrossRef]
8. Jacot, D.; Tosetti, N.; Pires, I.; Stock, J.; Graindorge, A.; Hung, Y.F.; Han, H.; Tewari, R.; Kursula, I.; Soldati-Favre, D. An Apicomplexan Actin-Binding Protein Serves as a Connector and Lipid Sensor to Coordinate Motility and Invasion. *Cell Host Microbe* **2016**, *20*, 731–743. [CrossRef]
9. Darvill, N.; Dubois, D.J.; Rouse, S.L.; Hammoudi, P.M.; Blake, T.; Benjamin, S.; Liu, B.; Soldati-Favre, D.; Matthews, S. Structural Basis of Phosphatidic Acid Sensing by APH in Apicomplexan Parasites. *Structure* **2018**, *26*, 1059–1071.e1056. [CrossRef]
10. Bullen, H.E.; Jia, Y.; Yamaryo-Botte, Y.; Bisio, H.; Zhang, O.; Jemelin, N.K.; Marq, J.B.; Carruthers, V.; Botte, C.Y.; Soldati-Favre, D. Phosphatidic Acid-Mediated Signaling Regulates Microneme Secretion in Toxoplasma. *Cell Host Microbe* **2016**, *19*, 349–360. [CrossRef]
11. Whitelaw, J.A.; Latorre-Barragan, F.; Gras, S.; Pall, G.S.; Leung, J.M.; Heaslip, A.; Egarter, S.; Andenmatten, N.; Nelson, S.R.; Warshaw, D.M.; et al. Surface attachment, promoted by the actomyosin system of Toxoplasma gondii is important for efficient gliding motility and invasion. *BMC Biol.* **2017**, *15*, 1–23. [CrossRef]
12. Beilsten-Edmands, J.; Winter, G.; Gildea, R.; Parkhurst, J.; Waterman, D.; Evans, G. Scaling diffraction data in the DIALS software package: Algorithms and new approaches for multi-crystal scaling. *Acta Crystallogr. D Struct. Biol.* **2020**, *76*, 385–399. [CrossRef]
13. Winn, M.D.; Ballard, C.C.; Cowtan, K.D.; Dodson, E.J.; Emsley, P.; Evans, P.R.; Keegan, R.M.; Krissinel, E.B.; Leslie, A.G.; McCoy, A.; et al. Overview of the CCP4 suite and current developments. *Acta Crystallogr. D Biol. Crystallogr.* **2011**, *67*, 235–242. [CrossRef]
14. Winter, G. xia2: An expert system for macromolecular crystallography data reduction. *J. Appl. Crystallogr.* **2010**, *43*, 186–190. [CrossRef]
15. Winter, G.; Waterman, D.G.; Parkhurst, J.M.; Brewster, A.S.; Gildea, R.J.; Gerstel, M.; Fuentes-Montero, L.; Vollmar, M.; Michels-Clark, T.; Young, I.D.; et al. DIALS: Implementation and evaluation of a new integration package. *Acta Crystallogr. D Struct. Biol.* **2018**, *74*, 85–97. [CrossRef] [PubMed]
16. Evans, P. Scaling and assessment of data quality. *Acta Crystallogr. D Biol. Crystallogr.* **2006**, *62*, 72–82. [CrossRef] [PubMed]
17. Winter, G.; Lobley, C.M.; Prince, S.M. Decision making in xia2. *Acta Crystallogr. D Biol. Crystallogr.* **2013**, *69*, 1260–1273. [CrossRef]
18. Vonrhein, C.; Flensburg, C.; Keller, P.; Sharff, A.; Smart, O.; Paciorek, W.; Womack, T.; Bricogne, G. Data processing and analysis with the autoPROC toolbox. *Acta Crystallogr. D Biol. Crystallogr.* **2011**, *67*, 293–302. [CrossRef]
19. Vonrhein, C.; Blanc, E.; Roversi, P.; Bricogne, G. Automated structure solution with autoSHARP. *Methods Mol. Biol.* **2007**, *364*, 215–230. [CrossRef]
20. Emsley, P.; Lohkamp, B.; Scott, W.G.; Cowtan, K. Features and development of Coot. *Acta Crystallogr. D Biol. Crystallogr.* **2010**, *66*, 486–501. [CrossRef] [PubMed]
21. Adams, P.D.; Afonine, P.V.; Bunkoczi, G.; Chen, V.B.; Davis, I.W.; Echols, N.; Headd, J.J.; Hung, L.-W.; Kapral, G.J.; Grosse-Kunstleve, R.W.; et al. PHENIX: A comprehensive Python-based system for macromolecular structure solution. *Acta Crystallogr. D Biol. Crystallogr.* **2010**, *66*, 213–221. [CrossRef] [PubMed]
22. Matthews, B.W. Solvent content of protein crystals. *J. Mol. Biol.* **1968**, *33*, 491–497. [CrossRef]
23. Senior, A.W.; Evans, R.; Jumper, J.; Kirkpatrick, J.; Sifre, L.; Green, T.; Qin, C.; Žídek, A.; Nelson, A.W.R.; Bridgland, A.; et al. Improved protein structure prediction using potentials from deep learning. *Nature* **2020**, *577*, 706–710. [CrossRef]
24. Källberg, M.; Wang, H.; Wang, S.; Peng, J.; Wang, Z.; Lu, H.; Xu, J. Template-based protein structure modeling using the RaptorX web server. *Nat. Protoc.* **2012**, *7*, 1511–1522. [CrossRef] [PubMed]
25. Yang, J.; Yan, R.; Roy, A.; Xu, D.; Poisson, J.; Zhang, Y. The I-TASSER Suite: Protein structure and function prediction. *Nat. Methods* **2015**, *12*, 7–8. [CrossRef]

MDPI

Communication

Sequence Analysis and Preliminary X-ray Crystallographic Analysis of an Acetylesterase (*Lg*EstI) from *Lactococcus garvieae*

Hackwon Do [1,2,†], Ying Wang [3,†], Chang Woo Lee [1], Wanki Yoo [4], Sangeun Jeon [3], Jisub Hwang [1,2], Min Ju Lee [1], Kyeong Kyu Kim [4], Han-Woo Kim [1,2], Jun Hyuck Lee [1,2,*] and T. Doohun Kim [3,*]

1 Research Unit of Cryogenic Novel Material, Korea Polar Research Institute, Incheon 21990, Korea; hackwondo@kopri.re.kr (H.D.); lcw@xenohelix.com (C.W.L.); hjsub9696@kopri.re.kr (J.H.); minju0949@kopri.re.kr (M.J.L.); hwkim@kopri.re.kr (H.-W.K.)
2 Department of Polar Sciences, University of Science and Technology, Incheon 21990, Korea
3 Department of Chemistry, Graduate School of General Studies, Sookmyung Women's University, Seoul 04310, Korea; ying.wang@oci.uni-hannover.de (Y.W.); sangeun94@sookmyung.ac.kr (S.J.)
4 Department of Molecular Cell Biology, Samsung Biomedical Research Institute, Sungkyunkwan University School of Medicine, Suwon 16419, Korea; vlqkshqk61@skku.edu (W.Y.); kyeongkyu@skku.edu (K.K.K.)
* Correspondence: junhyucklee@kopri.re.kr (J.H.L.); doohunkim@sookmyung.ac.kr (T.D.K.)
† These authors contributed equally to this work.

Abstract: A gene encoding *Lg*EstI was cloned from a bacterial fish pathogen, *Lactococcus garvieae*. Sequence and bioinformatic analysis revealed that *Lg*EstI is close to the acetyl esterase family and had maximum similarity to a hydrolase (UniProt: Q5UQ83) from *Acanthamoeba polyphaga mimivirus* (APMV). Here, we present the results of *Lg*EstI overexpression and purification, and its preliminary X-ray crystallographic analysis. The wild-type *Lg*EstI protein was overexpressed in *Escherichia coli*, and its enzymatic activity was tested using *p*-nitrophenyl of varying lengths. *Lg*EstI protein exhibited higher esterase activity toward *p*-nitrophenyl acetate. To better understand the mechanism underlying *Lg*EstI activity and subject it to protein engineering, we determined the high-resolution crystal structure of *Lg*EstI. First, the wild-type *Lg*EstI protein was crystallized in 0.1 M Tris-HCl buffer (pH 7.1), 0.2 M calcium acetate hydrate, and 19% (*w*/*v*) PEG 3000, and the native X-ray diffraction dataset was collected up to 2.0 Å resolution. The crystal structure was successfully determined using a molecular replacement method, and structure refinement and model building are underway. The upcoming complete structural information of *Lg*EstI may elucidate the substrate-binding mechanism and provide novel strategies for subjecting *Lg*EstI to protein engineering.

Keywords: esterase; *Lactococcus garvieae*; X-ray crystallography

Citation: Do, H.; Wang, Y.; Lee, C.W.; Yoo, W.; Jeon, S.; Hwang, J.; Lee, M.J.; Kim, K.K.; Kim, H.-W.; Lee, J.H.; et al. Sequence Analysis and Preliminary X-ray Crystallographic Analysis of an Acetylesterase (*Lg*EstI) from *Lactococcus garvieae*. *Crystals* **2022**, *12*, 46. https://doi.org/10.3390/cryst12010046

Academic Editor: Abel Moreno

Received: 16 December 2021
Accepted: 26 December 2021
Published: 29 December 2021

1. Introduction

Esterases (E.C. 3.1.1.X) catalyze the hydrolysis of various substrates containing ester groups. Esterases are serine hydrolases that contain a conserved Ser-Asp/Glu-His catalytic triad with an α/β hydrolase fold. Although esterases harbor the same α/β hydrolase fold and have high sequence homology, they have different substrate specificities and perform varying biological functions. Recently, esterases of microbial origin have gained significant interest in scientific research because of their potential applications in the biotechnology industry [1,2]. Therefore, extensive efforts are being made to identify unique esterases with higher activity, improved stability, and broad substrate specificity from newly found microorganism genomes as well as metagenomes. Such esterases can be further subjected to protein engineering to generate esterases with a precise structure and desirable functions [3].

To date, many microbial esterases have been discovered and characterized [4,5]. In particular, pathogenic bacteria have been well-known for producing various extracellular esterases. These secreted esterases may be important for the virulence and pathogenesis of some bacteria. In addition, some pathogenic bacterial esterases are involved in the drug resistance mechanism by enzymatic cleavage of antibiotics [6,7]. However, detailed enzymatic activity differences and the structural information of pathogenic bacterial esterases are still unclear. Recently, the complete genome sequence information of a major fish pathogen bacteria, *Lactococcus garvieae*, has been published. *Lactococcus garvieae* is a Gram-positive bacterium with a sphere-shape (cocci) [8].

In this study, several putative esterase encoding genes of *Lactococcus garvieae* were identified as targets and sequentially tested. In this study, we cloned the *Lg*EstI of *L. garvieae* into a plasmid vector to overexpress the *Lg*EstI protein in *Escherichia coli*. To obtain structural information, we crystallized *Lg*EstI to perform initial X-ray crystallographic experiments and successfully determined its structure. Further structural refinements and model building are underway. We believe that structural analysis of *Lg*EstI in the near future will add further value to our biochemical analysis and facilitate a better understanding of the potential application of *Lg*EstI, as well as provide new insights for further engineering of this protein.

2. Materials and Methods

2.1. Protein Clustering

ProtBLAST/PSI-BLAST was used to detect the remote homologs [9]. Initially, the *Lg*EstI sequence was blasted against the Protein Data Bank (PDB) database, and the result (E-value cutoff for reporting = 1×10^{-10}) was reloaded for performing a second blast against the Uniprot_sport database. Finally, the data with full-length sequences were forwarded to CLANS [10] for clustering by sequence similarity. The data were visualized with connecting lines, and shorter lines indicated higher sequence similarity.

2.2. Esterase Activity

p-Nitrophenyl esters were purchased from Sigma-Aldrich (Sigma, St. Louis, MO, USA) and used as substrates to assay *Lg*EstI activity. *Lg*EstI activity with different lengths of acyl carbon chains was determined by monitoring the production of *p*-nitrophenol (PNP) spectrophotometrically as previously described, with minor modifications [11].

2.3. Gene Cloning, Expression, and Purification of Recombinant LgEstI Protein

Genomic DNA of *L. garvieae* was isolated using a DNA extraction kit according to the manufacturer's instructions (Bioneer, Daejeon, Korea). The CDS of *Lg*EstI (NCBI accession number: WP_042219410.1) was amplified by performing PCR using appropriate primers. The PCR product was cloned into the pET21a vector between the *Nde*I and *Xho*I restriction sites. For protein expression, cells of the *E. coli* BL21 (DE3) strain were transformed with the plasmid harboring *Lg*EstI (Table 1). Fresh transformants were grown at 37 °C in 4 L of Luria–Bertani medium containing 50 μg/mL ampicillin. When the $O.D._{600}$ reached 0.5, protein overexpression was induced by adding 1.0 mM isopropyl β-D-1-thiogalactopyranoside (IPTG) to the cell culture, which was then continued for 20 h at 20 °C. Next, cells were pelleted by 20 min of centrifugation at 4000 rpm, resuspended in a lysis buffer (NPI-20, 50 mM NaH_2PO_4, 300 mM NaCl, and 20 mM imidazole), and disrupted by ultrasonication at 30% amplitude. An ice bath was used to maintain the temperature below 40 °C. The cell debris was removed by 1 h of centrifugation at 16,000 rpm and 4 °C, and the supernatant was used for the purification step.

The recombinant *Lg*EstI was purified using gravity-flow purification as the initial step. To eliminate the remaining free nickel, a pre-packed column with His-tagged resin was washed with 5 mL of NPI-20. The supernatant containing recombinant *Lg*EstI was slowly loaded onto the column and then washed by applying 10 CV with NPI-20. Next, the purified recombinant *Lg*EstI was eluted with NPI-300 (50 mM NaH_2PO_4, 300 mM NaCl, and 300 mM imidazole). The elution was concentrated to 5 mL and incubated with thrombin for 2 days in a rotating incubator at 4 °C to cleave the His tag. For the second purification, the tag-removed sample was loaded onto a Superdex prep grade column

(HiLoad® 16/600 Superdex® 200 pg) equilibrated with 20 mM Tris-HCl (pH 8.0), 200 mM NaCl, and 1 mM TCEP (Tris(2-carboxyethyl)phosphine hydrochloride) for SDS-PAGE. The Bradford protein assay was performed to confirm the purity and concentration of the recombinant *Lg*EstI.

Table 1. Recombinant *Lg*EstI protein production information.

***Lg*Est**	
Source organism	*Lactococcus garvieae*
DNA source	Genomic DNA
Cloning vector	pET21a
Expression host	*Escherichia coli* BL21 (DE3)
Amino acid sequence	MVERISLEKAALEFSEANAPHPRIYELPVEEGRSLLNEV QDSPVVKEDVDIEDIAVDTGEWGEINVRFIRPLHQEKKL PVIFYIHGAGWVFGNAHTHDKLIRELAVRTNSVVVFSE YSLSPEAKYPTAIEQNYAVLQQLKDFANDKKFDVNHLT VAGDSVGGNMATVMTLLTKQRGGQKIGQQVLYYPVT DANFDTDSYNEFAENYFLTKEGMIWFWDQYTTSQEER HQITASPLRATKEDLADLPAALIITGEADVLRDEGEAYA RKLREADVEVTQVRFQAIIHDFVMVNSMNETHATRAA MSLSTQWINEKNRK

2.4. LgEstI Crystallization, Data Collection, and Phasing

More than 1000 different crystallization screening solutions were screened for the initial crystallization trial using recombinant *Lg*EstI (25 mg/mL) by the sitting-drop vapor diffusion method. The crystallization conditions established from the initial screening were further optimized to generate diffraction-quality crystals (Table 2). Diffraction data were collected at the 5C beamline of the PLS, Korea. A single square pillar crystal was mounted on a goniometer equipped with a nitrogen stream with perfluoropolyether cryo-oil (Hampton Research, Laguna, CA, USA) as a cryoprotectant, with an oscillation range of 1°. Native data diffracting to 2.0 Å were collected, and processing and reduction were performed using XDS [12]. The *Lg*EstI phase was obtained by molecular replacement using the I-TASSER model [13]. Next, the coordinates of *Lg*EstI were built by iterative model building using a combination of Coot and Autobuild. Refinement of *Lg*EstI was performed using phenix.refine from Phenix. The details of selected X-ray data collection statistics are listed in Table 3.

Table 2. Crystallization details.

Method	**Vapor Diffusion**
Plate type for screening	96-well sitting drop MRC plate (Molecular Dimension, Suffolk, UK)
Plate type for optimization	24-well hanging drop plate (Molecular Dimension, Suffolk, UK)
Temperature (°C)	22
Protein concentration (mg/mL)	25
Composition of protein solution	20 mM Tris-HCl (pH 8.0), 200 mM NaCl, and 1 mM TECP
Composition of reservoir solution	0.1 M Tris: HCl (pH 7.1), 0.2 M calcium acetate hydrate, and 19% (*w*/*v*) PEG 3000
Volume and ratio of drop	2 µL, 1:1
Volume of reservoir (µL)	500

Table 3. X-ray diffraction data collection statistics.

Data Collection	*Lg*EstI
Wavelength (Å)	0.97949
X-ray source	BL-5C beamline
Rotation range per image (°)	1
Exposure time (s)	0.1
Space group	$I2_1$
Unit-cell parameters (Å, °)	a = 54.41, b = 92.77, c = 218.11, α = γ = 90.0, β = 96.61
Resolution range (Å) [a]	28.49–2.0 (2.05–2.0)
No. of observed reflections [a]	494,404
No. of unique reflections [a]	71,847 (4652)
Completeness (%) [a]	98.7 (99.8)
Redundancy [a]	6.9 (7.2)
R_{sym} [a,b]	0.059 (0.540)
I/σ [a]	16.1 (2.8)
Mosaicity	0.36
$CC_{\frac{1}{2}}$ [a,c]	(99.7%)

[a] Values in parentheses correspond to the highest resolution shell. [b] $R_{sym} = \Sigma_h\Sigma_i \mid I(h)_i - \langle I(h)\rangle \mid / \Sigma_h\Sigma_i I(h)_i$, where I is the intensity of the reflection h, Σ_h is the sum over all reflections, and Σ_i is the sum over i measurements of reflection h. [c] Percentage correlation between intensities from random half-datasets [14].

3. Results

3.1. Clustering Analysis of LgEstI

Clustering analysis of *Lg*EstI was performed against the PDB and UniProt databases for primary and secondary blasting, respectively, using the MPI Bioinformatics Toolkit. Available online: https://toolkit.tuebingen.mpg.de (accessed on 6 December 2021) [9]. In total, 170 proteins were detected and aligned to visualize these relationships. Four main clusters were identified from the clustering analysis. The acetyl esterase family is tightly clustered and shows a high degree of sequence similarity. The second cluster included a combination of carboxylesterase, acetylcholinesterase, and para-nitrobenzyl esterase. Arylacetamide deacetylases and tuliposide A-converting enzymes were grouped into separate clusters. Interestingly, *Lg*EstI could not be assigned to any of these clusters and showed maximum similarity to a hydrolase (UniProt: Q5UQ83) of *Acanthamoeba polyphaga mimivirus* (APMV) within the UniProt database. Based on the length of the connection and the color indicating sequence similarity, *Lg*EstI appeared close to the acetyl esterase group but was not part of it. The clustering analysis confirmed that *Lg*EstI has a unique sequence and may have a distinctive function from the other enzyme clusters (Figure 1).

3.2. Purification and Biochemical Characterization of LgEstI

To conduct a preliminary study for understanding the detailed functional mechanisms of *Lg*EstI, we cloned the *Lg*EstI of *L. garvieae* into a plasmid vector and overexpressed the *Lg*EstI protein in *E. coli*. Purified *Lg*EstI protein was used to perform preliminary X-ray crystallography. Two conventional purification steps, namely tag-affinity purification and size-exclusion chromatography, were applied to purify the recombinant *Lg*EstI protein (>95%), and the final protein purity was checked on SDS-PAGE. The molecular weight of *Lg*EstI was approximately 37 kDa, which is similar to the calculated molecular weight based on the amino acid sequence (Figure 2A). The enzyme activity assay using purified *Lg*EstI protein with various *p*-nitrophenyl esters showed that *Lg*EstI had a narrow substrate specificity, as it was active only against *p*-nitrophenyl acetate (Figure 2B). Thus, the clustering and enzyme activity in combination indicated that *Lg*EstI exhibits specific activity against analogs of acetate.

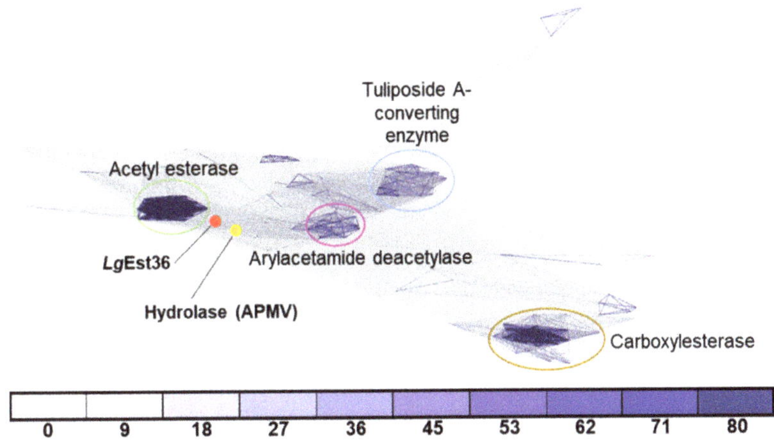

Figure 1. Sequences of orthologs from different species were used for clustering by CLANS clustering. ProtBLAST/PSI-BLAST was used to detect distantly related proteins using the Protein Data Bank (PDB) and Uniprot_sport databases. *Lg*EstI and the hydrolase (UniProt: Q5UQ83) from *Acanthamoeba polyphaga mimivirus* (APMV) within the UniProt database are indicated by red and yellow dots, respectively. Connecting lines of darker intensity and shorter length indicate higher sequence similarity. Connections with a *p*-value better than 1×10^{-x} are drawn in the corresponding color (x = number below).

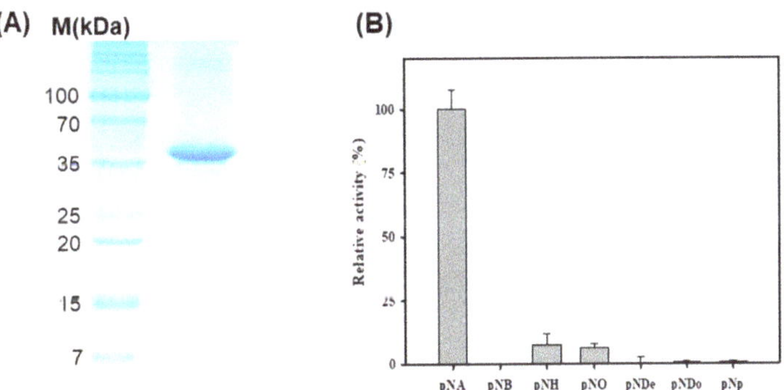

Figure 2. Purification and activity analysis of *Lg*EstI using various *p*-nitrophenyl esters. (**A**) SDS-PAGE analysis of *Lg*EstI purified by tag-affinity purification and size-exclusion chromatography. (**B**) The relative activities of *Lg*EstI against *p*-nitrophenyl esters, expressed as percentages. This activity assay was carried out using various substrates (*p*NA: *p*-nitrophenyl acetate; *p*NB: *p*-nitrophenyl butyrate; *p*NH: *p*-nitrophenyl hexanoate; *p*NO: *p*-nitrophenyl octanoate; *p*NDe: *p*-nitrophenyl decanoate; *p*NDo: *p*-nitrophenyl dodecanoate; *p*NP: *p*-nitrophenyl phosphate). The activity of *Lg*EstI with *p*-nitrophenyl acetate was defined as 100%.

3.3. X-ray Crystallographic Study of *Lg*EstI

To obtain the factual structural information of *Lg*EstI, we performed crystallization and initial X-ray crystallographic experiments. The crystallization conditions for *Lg*EstI protein were screened using more than 1000 different crystallization buffers. After optimizing the crystallization conditions by changing the precipitant concentration and pH in the reservoir solution and drops, rhombus-shaped crystals (approximately 100×200 μm) were obtained using 0.1 M Tris-HCl (pH 7.1), 0.2 M calcium acetate hydrate, and 19% (*w*/*v*) PEG 3000 (Figure 3A). However, X-ray diffraction of *Lg*EstI crystals resulted in a poor diffraction pattern with a resolution of approximately 4 Å. Instead of searching for new crystallization conditions where *Lg*EstI could be better diffracted, we aimed to optimize the cryoprotectant conditions. *Lg*EstI crystals soaked in solutions containing perfluoropolyether cryo-oil (Hampton Research, Laguna, CA, USA) for 10 s exhibited a dramatic improvement in diffraction quality (~2 Å), whereas the crystals cryoprotected with general cryoprotectants (e.g., glycerol and PEG) showed low-resolution diffraction patterns (Figure 3B). Briefly, we tested more than 50 *Lg*EstI crystals to find the optimal cryoprotectant solution. *Lg*EstI crystals were very unstable in glycerol or PEG-containing cryoprotectant solutions, resulting in the *Lg*EstI crystals melting and cracking. However, *Lg*EstI crystals were more stable for a long time (more than 5 min) without melting or cracking in oil-based cryoprotectant solutions (Paratone-*N* oil and perfluoropolyether cryo-oil). We used the same conditions and similar sized crystals grown under the same crystallization conditions for this experiment. Finally, we obtained the best *Lg*EstI crystals for X-ray diffraction data collection under the conditions with perfluoropolyether cryo-oil. During X-ray diffraction data collection, flash cooling the crystal in a nitrogen gas stream at 100 K prevented crystal cracking. We inferred that an oil-based solution could be a suitable cryoprotectant for *Lg*EstI. All the diffraction data (completeness: 98.7%) were collected at a resolution of 2.0 Å. The data processing program XDS was used to index, integrate, and scale the acquired diffraction data [12].

(A)

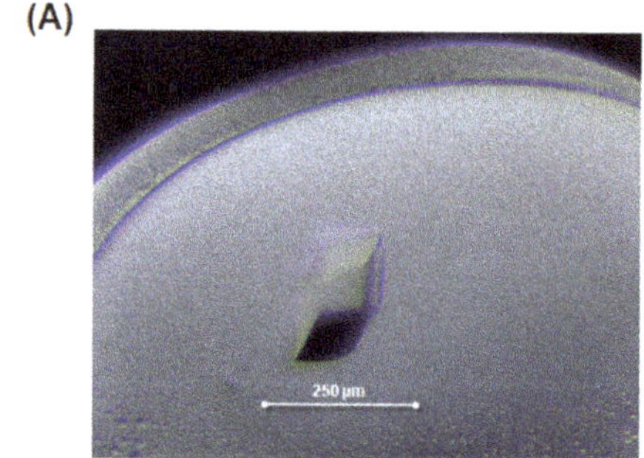

(B)

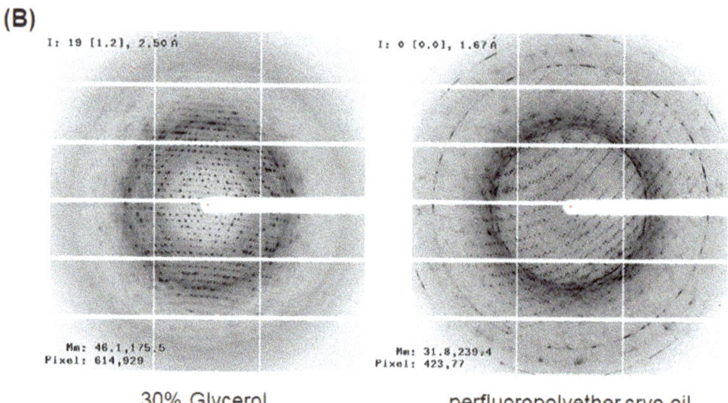

Figure 3. X-ray crystallographic study of *Lg*EstI. (**A**) Crystal of *Lg*EstI protein grown in 0.1 M Tris-HCl (pH 7.1), 0.2 M calcium acetate hydrate, and 19% (*w*/*v*) PEG 3000. (**B**) Cryoprotectant optimization for the X-ray diffraction study. *Lg*EstI crystals soaked in solutions containing perfluoropolyether cryo-oil for 10 s and flash-frozen under a nitrogen stream are shown (**right**).

The initial structure of *Lg*EstI was successfully determined. For phasing *Lg*EstI, we first generated the *Lg*EstI model based on the homologous structures using the protein structure prediction server I-TASSER [13]. The quality of the model was confirmed to have a C-score of 1.40 and a TM-score of 0.91 $\pm$ 0.06, respectively. Next, the *Lg*EstI model was used to overcome the phase problem using the molecular replacement method. Further structural refinements and model building are underway in our laboratory. We believe that a structural analysis of *Lg*EstI in the near future will facilitate a better understanding of our biochemical analysis results, as well as providing valuable insights for further engineering of *Lg*EstI protein.

Author Contributions: Conceptualization, J.H.L. and T.D.K.; methodology, H.D., Y.W. and C.W.L.; validation, W.Y., S.J., J.H., M.J.L. and H.-W.K.; resources, M.J.L. and H.-W.K.; writing—original draft preparation, H.D., Y.W. and K.K.K.; writing—review and editing, J.H.L. and T.D.K.; funding acquisition, H.-W.K., J.H.L. and T.D.K. All authors have read and agreed to the published version of the manuscript.

Funding: This work was financially supported by an academic grant (NRF-2021R1F1A1048135) from the National Research Foundation of Korea (T.D.K). This research was also supported by a National Research Foundation of Korea Grant from the Korean Government (MSIT; the Ministry of Science and ICT), NRF-2021M1A5A1075524) (KOPRI-PN21014) (H.-W.K). This research was a part of the project titled "Development of potential antibiotic compounds using polar organism resources (15250103, KOPRI Grant PM21030)", funded by the Ministry of Oceans and Fisheries, Korea (J.H.L).

Institutional Review Board Statement: Not applicable.

Informed Consent Statement: Not applicable.

Data Availability Statement: Not applicable.

Conflicts of Interest: The authors declare no conflict of interest.

References

1. Fazary, A.E.; Ju, Y.H. Feruloyl Esterases as Biotechnological Tools: Current and Future Perspectives. *Acta Biochim. Biophys. Sin.* **2007**, *39*, 811–828. [CrossRef] [PubMed]
2. Celligoi, M.A.P.C.; Baldo, C.; De Melo, M.R.; Gasparin, F.G.M.; Marques, T.A.; De Barros, M. Lipase properties, functions and food applications, microb. In *Microbial Enzyme Technology in Food Applications*; CRC Press: Boca Raton, FL, USA, 2017; pp. 214–240. [CrossRef]
3. Pleiss, J.; Fischer, M.; Peiker, M.; Thiele, C.; Schmid, R.D. Lipase Engineering Database. *J. Mol. Catal. B Enzym.* **2000**, *10*, 491–508. [CrossRef]
4. Bornscheuer, U.T. Microbial carboxyl esterases: Classification, properties and application in biocatalysis. *FEMS Microbiol. Rev.* **2002**, *26*, 73–81. [CrossRef] [PubMed]
5. Barzkar, N.; Sohail, M.; Tamadoni Jahromi, S.; Gozari, M.; Poormozaffar, S.; Nahavandi, R.; Hafezieh, M. Marine Bacterial Esterases: Emerging Biocatalysts for Industrial Applications. *Appl. Biochem. Biotechnol.* **2021**, *193*, 1187–1214. [CrossRef] [PubMed]
6. Larsen, E.M.; Johnson, R.J. Microbial esterases and ester prodrugs: An unlikely marriage for combating antibiotic resistance. *Drug Dev. Res.* **2019**, *80*, 33–47. [CrossRef] [PubMed]
7. Kwon, S.; Yoo, W.; Kim, Y.O.; Kim, K.K.; Kim, T.D. Molecular Characterization of a Novel Family VIII Esterase with β-Lactamase Activity (PsEstA) from *Paenibacillus* sp. *Biomolecules* **2019**, *9*, 786. [CrossRef] [PubMed]
8. Morita, H.; Toh, H.; Oshima, K.; Yoshizaki, M.; Kawanishi, M.; Nakaya, K.; Suzuki, T.; Miyauchi, E.; Ishii, Y.; Tanabe, S.; et al. Complete genome sequence and comparative analysis of the fish pathogen *Lactococcus garvieae*. *PLoS ONE* **2011**, *6*, e23184. [CrossRef] [PubMed]
9. Gabler, F.; Nam, S.Z.; Till, S.; Mirdita, M.; Steinegger, M.; Söding, J.; Lupas, A.N.; Alva, V. Protein Sequence Analysis Using the MPI Bioinformatics Toolkit. *Curr. Protoc. Bioinform.* **2020**, *72*, e108. [CrossRef] [PubMed]
10. Frickey, T.; Lupas, A. CLANS: A Java Application for Visualizing Protein Families Based on Pairwise Similarity. *Bioinformatics* **2004**, *20*, 3702–3704. [CrossRef] [PubMed]
11. Lee, C.W.; Kwon, S.; Park, S.H.; Kim, B.Y.; Yoo, W.; Ryu, B.H.; Kim, H.W.; Shin, S.C.; Kim, S.; Park, H.; et al. Crystal Structure and Functional Characterization of an Esterase (EaEST) from Exiguobacterium antarcticum. *PLoS ONE* **2017**, *12*, e0169540. [CrossRef] [PubMed]
12. Kabsch, W. XDS. *Acta Crystallogr. D Biol. Crystallogr.* **2010**, *66*, 125–132. [CrossRef] [PubMed]

13. Yang, J.; Yan, R.; Roy, A.; Xu, D.; Poisson, J.; Zhang, Y. The I-TASSER Suite: Protein Structure and Function Prediction. *Nat. Methods* **2015**, *12*, 7–8. [CrossRef] [PubMed]
14. Karplus, P.A.; Diederichs, K. Linking Crystallographic Model and Data Quality. *Science* **2012**, *336*, 1030–1033. [CrossRef] [PubMed]

MDPI

Communication

Structural and Biochemical Studies of *Bacillus subtilis* MobB

Dajeong Kim [1], Sarah Choi [1,2], Hyunjin Kim [1] and Jungwoo Choe [1,*]

[1] Department of Life Science, University of Seoul, Seoul 02504, Korea; da1233@khu.ac.kr (D.K.); sarahchoi@uos.ac.kr (S.C.); hkim@uos.ac.kr (H.K.)
[2] Macrogen Inc., Seoul 08511, Korea
* Correspondence: jchoe21@uos.ac.kr; Tel.: +82-2-6490-2673

Abstract: The biosynthesis of molybdenum cofactor for redox enzymes is carried out by multiple enzymes in bacteria including MobA and MobB. MobA is known to catalyze the attachment of GMP to molybdopterin to form molybdopterin guanine dinucleotide. MobB is a GTP binding protein that enhances the activity of MobA by forming the MobA:MobB complex. However, the mechanism of activity enhancement by MobB is not well understood. The structure of *Bacillus subtilis* MobB was determined to 2.4 Å resolution and it showed an elongated homodimer with an extended β-sheet. Bound sulfate ions were observed in the Walker A motifs, indicating a possible phosphate-binding site for GTP molecules. The binding assay showed that the affinity between *B. subtilis* MobA and MobB increased in the presence of GTP, suggesting a possible role of MobB as an enhancer of MobA activity.

Keywords: molybdenum cofactor; MobB; Walker A motif; *Bacillus subtilis*; crystallography

Citation: Kim, D.; Choi, S.; Kim, H.; Choe, J. Structural and Biochemical Studies of *Bacillus subtilis* MobB. *Crystals* **2021**, *11*, 1262. https://doi.org/10.3390/cryst11101262

Academic Editors: Kyeong Kyu Kim and T. Doohun Kim

Received: 14 September 2021
Accepted: 15 October 2021
Published: 18 October 2021

1. Introduction

Molybdenum is an essential trace element required in diverse redox reactions in bacteria and eukaryotes [1]. A basic molybdenum cofactor (Moco) is a form in which the molybdenum atom is coordinated to the dithiolate moiety of a tricyclic pterin, called Molybdopterin (MPT) [2]. Moco biosynthesis is carried out by a conserved pathway with multiple steps (Figure 1): first, the formation of cyclic pyranopterin monophosphate from guanosine triphosphate (GTP) [3], followed by the insertion of two sulfur atoms to form MPT [4,5], and then the addition of a molybdenum atom to form molybdenum cofactor (Moco) via MPT-AMP intermediates [6–10]. In many bacteria, additional modifications of Moco occur by the attachment of guanosine monophosphate (GMP) or cytosine monophosphate (CMP) to form MPT guanine dinucleotide (MGD) [11] or MPT cytosine dinucleotide (MCD) [12], respectively. Two MGDs can be ligated to a single molybdenum atom, forming the bis-MGD cofactor, a reaction catalyzed by MobA and MobB proteins [13]. MobA is crucial for this reaction and MobB, a GTP-binding protein with weak intrinsic GTPase activity [14], enhances the function of MobA by forming the MobA:MobB complex as shown by an increased activity of nitrate reductase that requires bis-MGD as a cofactor [11]. *B. subtilis* MobB consists of 173 amino acids with a molecular weight of 19.5 KDa. Homologous structures of *B. subtilis* MobB include *Geobacillus stearothermophilus* (PDB ID: 1XJC), *Archaeoglobus fulgid* (2F1R), and *Escherichia coli* MobB (1NP6) with sequence identities of 40.6, 31.1 and 25.2% to *B. subtilis* MobB, respectively.

Our crystal structure of *B. subtilis* MobB showed that it formed an elongated homodimer. Each subunit contained a Walker A motif for binding the phosphate group of GTP, where a bound sulfate ion from crystallization solution was observed. It was shown that MobB interacts with MobA in vivo using a bacterial two-hybrid system [15], and the model of the MobA:MobB complex structure suggested that GTP was bound in the MobA:MobB interface [16]. We performed binding assays using *B. subtilis* MobA and MobB, which showed that these two proteins interacted more strongly in the presence of GTP in agreement with the suggested MobA:MobB complex model.

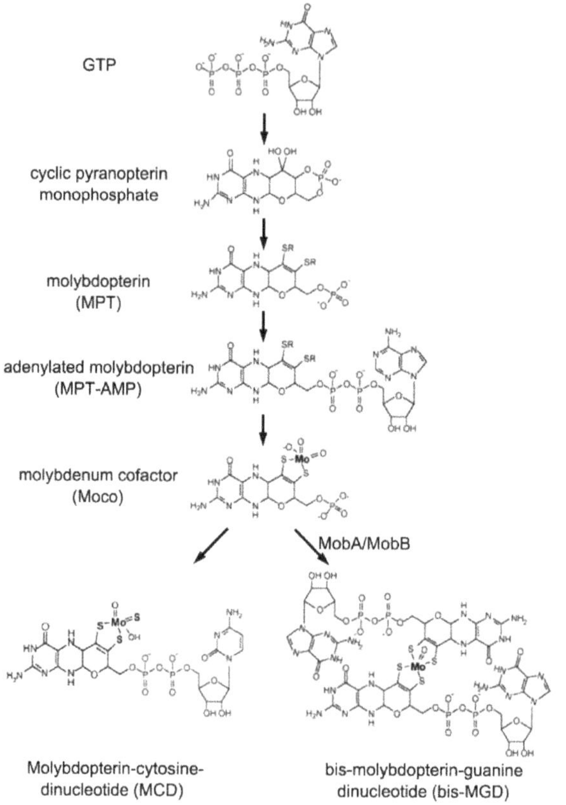

Figure 1. Biosynthesis of molybdenum cofactor (Moco) and further modifications to MCD and bis-MGD in bacteria. MobA and MobB catalyze the conversion of Moco to bis-MGD.

2. Materials and Methods

2.1. Cloning, Expression, and Purification

The *mobB* gene was amplified from *B. subtilis* genomic DNA by polymerase chain reaction (PCR) using primers (5′-GCTAGCATGGCCTTGGTCCGTCCTTTC-3′, 5′-GAATTCTTA TGCAGATTCCCCCTTCAGC-3′). The purified PCR product was cloned using NheI and EcoRI enzymes into the pET28b vector with an N-terminal His_6-tag and thrombin site. The construct was then transformed into BL21(DE3) *E. coli* strain (Novagen). The cells were cultured in LB media containing 30 μg/mL of kanamycin at 310 K until an OD_{600} of 0.6. The temperature was lowered to 291 K, followed by induction with 1 mM isopropyl β-D-1-thiogalactopyranoside (IPTG). Cell growth continued for 16 h, after which cells were harvested by centrifugation. Cell pellets were then resuspended in 20 mM Tris-HCl pH 7.5 and 250 mM NaCl buffer (lysis buffer) and lysed by sonication. The lysate was cleared by centrifugation, after which the supernatant was loaded onto a Ni-sepharose 6 affinity column (GE Healthcare) and eluted by a stepwise gradient of 50–800 mM imidazole in lysis buffer. After the N-terminal His_6-tag from the protein was cut by thrombin at 277 K for 16 h, MobB was further purified using a Superdex75 size-exclusion column (GE Healthcare) equilibrated with a buffer composed of 20 mM Tris-HCl pH7.5, 250 mM NaCl, 2 mM dithiothreitol (DTT), and 2 mM EDTA. The purity of the protein was analyzed by SDS-PAGE and the yield was about 2 mg from 1L culture. MobB proteins were concentrated by centrifugal ultrafiltration (Amicon, mol. wt. cutoff = 5 kDa) to 10 mg/mL as measured by Bradford assay (Thermo scientific).

2.2. Crystallization, Data Collection, and Structure Determination

Crystals of *B. subtilis* MobB were obtained by the hanging-drop vapor-diffusion method performed at 293 K by mixing 1 μL of protein with 1 μL of a well solution containing 23% (*w*/*v*) polyethyleneglycol (PEG) 3350 and 0.4 M ammonium sulfate. The protein used for crystallization is residues 1-173 of *B. subtilis* MobB (sequence ID: O31704) with amino acids "GSHMAS" left on the N-terminus from the vector. The crystals were rod-shaped and reached full size after 3 days. Crystals were transferred into a cryoprotectant solution composed of 25% (*w*/*v*) PEG 3350, 0.4 M ammonium sulfate, and 20% glycerol and then flash-cooled in liquid nitrogen. X-ray diffraction data were collected to 2.4 Å resolution at the PAL beamline 5C (KOREA). Data were processed using HKL-2000 [17] and the initial model was obtained by molecular replacement using the Phaser program [18] of the CCP4 package [19] with *G. stearothermophilus* MobB structure (1XJC) as a search model. The space group was C2 and the asymmetric unit contained five subunits in which subunit B formed a homodimer with subunit E, and C with D. Subunit A formed a homodimer with its 2-fold crystallographic symmetry-related molecule. The Matthews' coefficient (Vm) was 2.19 $Å^3$/Da, and the estimated solvent content was 43.8%. The model was refined with the Refmac [20,21] and PHENIX programs [22], and manual model building was performed using Coot software [23]. Data collection and refinement statistics are summarized in Table 1. Residues that were poorly observed in the electron density maps were not included in the final model (residues 1–6, 45–59, and 170–173 in A, C, and D subunits; residues 1–5, 45–59, and 173 in the B subunit; and residues 1–6, 45–59 and 172–173 in the E subunit). The Ramachandran plot produced by MolProbity [24] showed that 100% of residues were in the allowed or favored region. The coordinates and structure factors for *B. subtilis* MobB have been deposited in the RCSB Protein Data Bank with accession code 4OYH.

Table 1. Data Collection and Refinement Statistics.

Data Collection	
Space Group	C2
Unit Cell parameters (Å, °)	a = 225.53, b = 42.11, c = 93.62, β = 100.99
Wavelength (Å)	0.97921
Temperature (K)	100
Resolution (Å) [a]	50.0–2.40 (2.44–2.40)
No. of observed reflections	117,931
No. of unique reflections	33,530
Completeness (%)	98.8 (99.7)
Rsym [b]	0.109 (0.666)
<I/σ (I)> [c]	9.0 (3.1)
Refinement statistics	
No. of Residues	746
No. of water molecules	218
No. of sulfate ions	13
R_{cryst} (%), R_{free} (%) [d]	21.0, 30.3
rmsd bonds (Å)	0.014
rmsd angles (°)	1.559
Ramachandran plot	
Most favored region (%)	96.8
Additionally allowed region (%)	3.2

[a] Resolution range of the highest shell is listed in parentheses [b] $R_{sym} = \sum I - \langle I \rangle / \sum I$, where I is the intensity of an individual reflection and <I> is the average intensity over symmetry equivalents [c] <I/σ (I)> is the mean reflection intensity/estimated error [d] $R_{cryst} = \sum | |F_0| - |F_c| | / \sum |F_0|$, where F_o and F_c are the observed and calculated structure factor amplitudes, R_{free} is equivalent to R_{cryst} but calculated for a randomly chosen set of reflections that were omitted from the refinement process.

2.3. MobA Preparation and MobA:MobB Binding Assay

The *mobA* gene was amplified from *B. subtilis* genomic DNA by PCR using primers (5′-TACTTCCAATCCAATGCAATGAAGCATATAAA-TGTACTGCT-3′, 5′-TTATCCACTTCCA ATGTTATTATCAGTCCCACCTGAAGGAG-3′). The purified PCR product was cloned into a pLIC-Tr3Ta-HA vector containing an N-terminal His_6-tag and a TEV protease cleavage site. The construct was then transformed into BL21(DE3) *E. coli* strain (Novagen). MobA protein was expressed and purified using the same protocol as MobB described above. MobA:MobB binding assay was performed using the BLItz instrument (fortebio) and a Ni-NTA biosensor. BLItz system uses a bio-layer interferometry technology that can detect the change of the thickness of the coating on the biosensor that is proportional to the number of bound molecules [25]. BLItz system measures the association (k_{on}) and dissociation (k_{off}) rate constants to obtain K_D value (=k_{on}/k_{off}). All proteins were prepared to 2 mg/mL in 20 mM Tris-HCl pH 7.5 and 250 mM NaCl. MobA contained an N-terminal His_6-tag, whereas MobB did not contain a His_6-tag. After MobA with a His_6-tag was bound to the Ni-NTA sensor, the binding of MobB to MobA was monitored. Binding assays were performed in 20 mM Tris-HCl pH 7.5 and 250 mM NaCl $\pm$ 0.5 mM dGTP. Binding assay data were analyzed using BLItz Pro software version 1.1 to calculate K_D values.

3. Results and Discussions

The structure of *B. subtilis* MobB was determined at 2.4 Å resolution and refined to final R_{work} (R_{free}) values of 21.0% (30.3%). There were 5 subunits (two and a half dimers) in the asymmetric unit, and they had similar structures with RMSD's of 0.673, 0.991, 1.10, and 0.708 Å, respectively, compared to the A subunit when 148 Cα carbons are superposed. *B. subtilis* MobB formed a homodimer and each subunit consisted of six α-helices and eight β-strands (Figure 2A). *B. subtilis* MobB was eluted from the Superdex75 size-exclusion column (GE Healthcare) close to 60 kDa based on the calibration curve, and there were no peaks close to its monomer size (19.5 kDa) (Figure S1, in Supplementary Materials). Although this calculated MW of 60 kDa is closer to the trimeric form of MobB (58.5 kDa), we believe this peak corresponds to an elongated dimer of MobB, as observed in the crystal structure (Figure 2A). Interestingly, eight β-strands of subunit A and eight β-strands of subunit B (denoted with ′) formed a contiguous β-sheet composed of 16 β-strands. β-sheets at the ends (β8-β7-β6-β1 and β1′-β6′-β7′-β8′) were more twisted than the central part (β5-β2-β3′-β4′-β4-β3-β2′-β5′) of this contiguous β-sheet. Two molecules of MobB formed a closely intertwined homodimer with a dimer interface formed mainly by six β-strands (β2-β3′-β4′-β4-β3-β2′) in the center of the structure and α-helices (α2 and α3 from subunit A and α2′ and α3′ from subunit B) on both sides of the sheet (Figure 2A).

Analysis of the dimer interface using the PISA program [26] showed that the buried surface area was 2371 Å^2, which is about 21% of the total surface area of each monomer. There were 34 hydrogen bonds and three ionic interactions found in the dimer interface. Interactions from the inter-subunit β-sheet (β2-β3′, β4′-β4, and β3-β2′) contributed 18 hydrogen bonds to the interface. *B. subtilis* MobB showed similar structures to its homologs with known structures including *G. stearothermophilus* (PDB ID: 1XJC), *A. fulgid* (2F1R), and *E. coli* MobB (1NP6) [16], with RMSD's of 1.15, 2.39, and 3.32 Å, respectively (Figure S2, in Supplementary Materials). All of them formed a homodimer with a central β-sheet forming the major dimerization interface.

B. subtilis MobB protein contains a phosphate-binding site composed of a conserved Walker A motif (also known as the P-loop) [27] at the N-terminal region (GFQNSGKTT) (Figure 3C). A bound sulfate ion, used in the crystallization solution was observed in this motif in all subunits in the asymmetric unit (Figure 3A). The average B-factor of all sulfate ions is 75.1 Å^2 with occupancies set to 1. For comparison, the average B-factors of all protein atoms and water molecules are 54.8 and 69.6 Å^2, respectively. Oxygen atoms of the sulfate group were hydrogen-bonded to the amide nitrogen atoms of N20, S21, G22, and K23. The side chains of K23 and T24 also formed hydrogen bonds with the oxygen atoms of the sulfate group. Among these residues, G22, K23, and T24 were completely conserved in the multiple sequence alignments of MobB homologs (Figure 3C). Bound sulfate ions have been observed previously in the Walker A motif of *E. coli* MobB [16], where it was interacting with the amide groups of G16, G18, K19, and T20 (corresponding to N20, G22, K23, and T24 of *B. subtilis* MobB) in a similar manner to *B. subtilis* MobB. Another bound sulfate ion was observed in a nearby region, hydrogen-bonded to the amide groups of F111 and K112 of *B. subtilis* MobB, (Figure 3B) in all of the five subunits in the asymmetric unit. Although these two residues are identical in *E. coli* and *B. subtilis* MobB (Figure 3C, indicated by an asterisk), sulfate ion was not observed at this position in *E. coli* MobB, probably because the conformations of the amide groups are different from each other. Examination of the electrostatic potential surface of MobB dimer calculated by PyMOL [28] showed that both sulfate binding sites of *B. subtilis* MobB were predominantly positively charged (Figure 2B). Bound sulfate ions were not observed in the case of other homologs such as *A. fulgid* and *G. stearothermophilus* MobB either in the Walker A motif or the second sulfate binding site. This is probably because sulfate ion was not included in the crystallization condition.

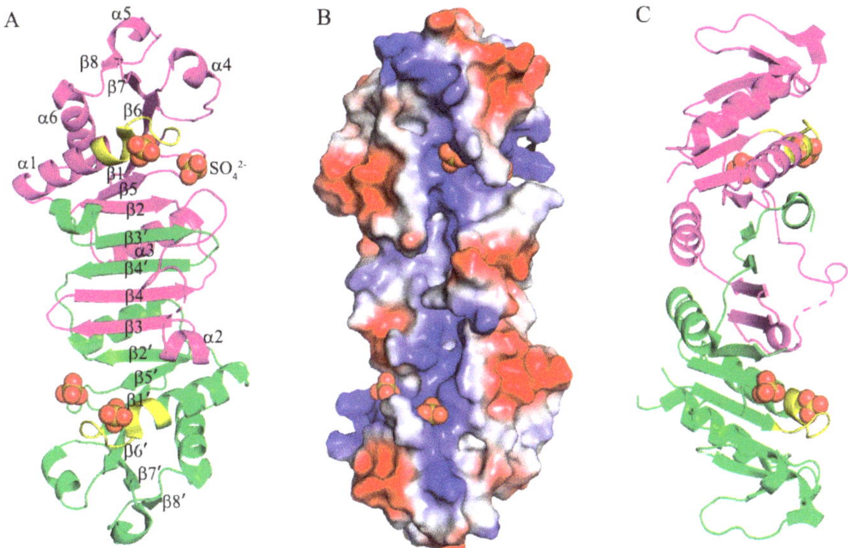

Figure 2. The overall structure of MobB dimer (**A**) Two subunits of MobB dimer (magenta and green) are drawn in cartoon representation and the secondary structures are labeled. Four bound sulfate ions are drawn in a ball-and-stick model. The Walker A motifs are highlighted in yellow in both subunits. (**B**) Electrostatic potential surface of MobB dimer as shown in (**A**) drawn in −75 to 75 K_bT/e_c. (**C**) 90° rotated view of (**A**).

MobA catalyzes the transfer of GMP to MPT to form MGD. This reaction is enhanced in the presence of MobB by forming a MobA:MobB complex [11]. We cloned the MobA gene from the *B. subtilis* genome and measured the binding affinity between MobA and MobB using the BLItz system. The dissociation constant (K_D) was 12.2 μM in the presence of 0.5 mM GTP and 27.3 μM in the absence of GTP (Figure 4). The model of MobA:MobB complex suggested that both MobA and MobB bind GTP [16] and the GTP-dependence of MobA:MobB binding affinity seems to support this model. We also observed that MobA precipitated quickly in the absence of MobB and became more stable when mixed with MobB during the purification process, suggesting the formation of a MobA:MobB complex.

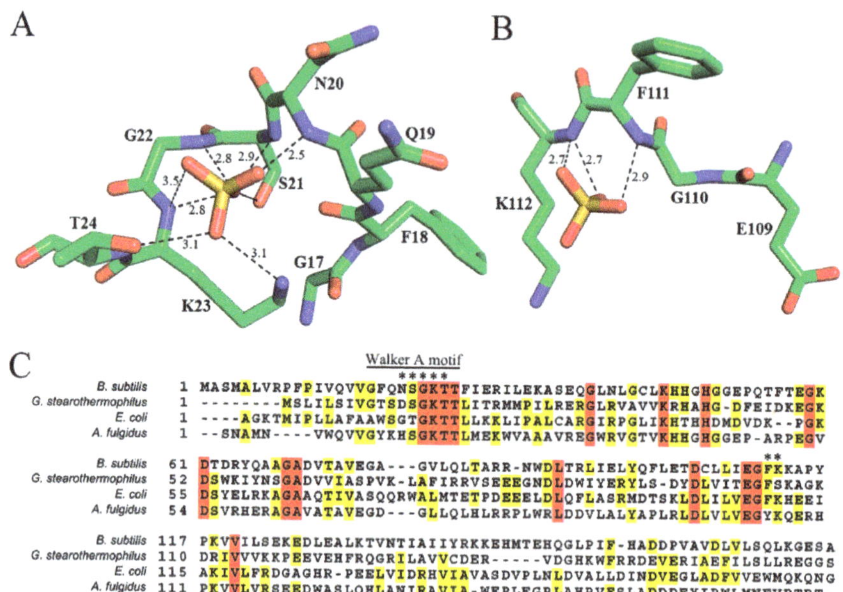

Figure 3. Sulfate-binding sites in MobB (**A**) bound sulfate ion in the Walker A motif is drawn in the ball-and-stick model. The hydrogen bonds are shown as dashed lines with the distances indicated in Å. (**B**) Sulfate ion in the second binding site. (**C**) Multiple sequence alignment of MobB homologs. Completely conserved residues are colored red and similar residues in yellow. Residues that are hydrogen-bonded with sulfate ions are marked with an asterisk.

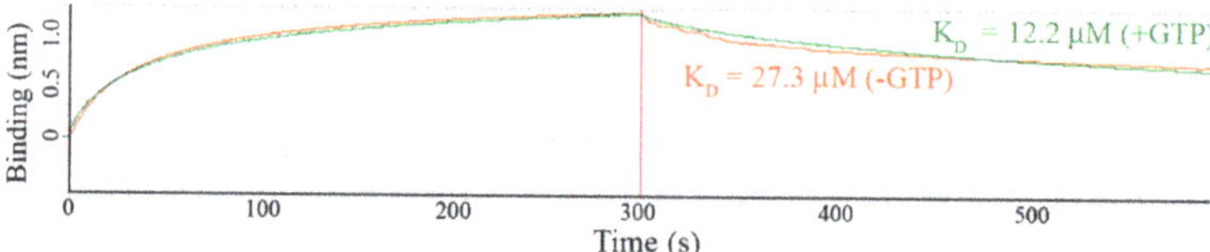

Figure 4. Binding assay between *B. subtilis* MobA and MobB. The binding affinity was measured using the BLItz system. Response refers to a change in the thickness of the layer on the biosensor. The raw data and calculated K_D values in the absence (orange) and presence (green) of GTP are shown.

In summary, the crystal structure of *B. subtilis* MobB showed that it formed a tightly bound homodimer with a bound sulfate ion in the Walker A motif. A second sulfate ion was observed in the vicinity of the first sulfate ion, which is unique to *B. subtilis* MobB. The binding affinity between MobA:MobB was increased in the presence of GTP, supporting the model that GTP is bound by both MobA and MobB during complex formation. We believe our structural and biochemical studies of MobB provide insight into the role of MobB as an enhancer of MobA by binding GTP molecule in the MobA and MobB interface. Future study of MobA:MobB complex structure with bound GTP will provide detailed information to understand the mechanism of MobB function.

Supplementary Materials: The following are available online at https://www.mdpi.com/article/10.3390/cryst11101262/s1, Figure S1: Superdex75 size-exclusion column profile of *B. subtilis* MobB (blue) and calibration standards (green). The molecular weights of the standard proteins are indicated in kDa; Figure S2: The structure of *B. subtilis* MobB (green) is superposed to (A) *G. stearothermophilus* (or-ange), (B) *A. fulgid* (blue), and (C) *E. coli* MobB (magenta).

Author Contributions: Conceptualization, J.C.; data acquisition and analysis, D.K., S.C. and H.K.; writing, D.K. and J.C.; project administration, J.C.; funding acquisition, J.C. All authors have read and agreed to the published version of the manuscript.

Funding: This research was funded by The National Research Foundation of Korea (NRF-2018R1D1A 1A09083579).

Institutional Review Board Statement: Not applicable.

Informed Consent Statement: Not applicable.

Data Availability Statement: Publicly available datasets were analyzed in this study. This data can be found here: [https://www.rcsb.org (PDB ID:4OYH)].

Acknowledgments: We thank the staff members of Pohang Synchrotron Laboratory (PAL) beamline 5C for data collection.

Conflicts of Interest: The authors declare no conflict of interest.

References

1. Hille, R. The mononuclear molybdenum enzymes. *Chem. Rev.* **1996**, *96*, 2757–2816. [CrossRef] [PubMed]
2. Rajagopalan, K.V.; Johnson, J.L. The pterin molybdenum cofactors. *J. Biol. Chem.* **1992**, *267*, 10199–10202. [CrossRef]
3. Hover, B.M.; Loksztejn, A.; Ribeiro, A.A.; Yokoyama, K. Identification of a Cyclic Nucleotide as a Cryptic Intermediate in Molybdenum Cofactor Biosynthesis. *J. Am. Chem. Soc.* **2013**, *135*, 7019–7032. [CrossRef] [PubMed]
4. Gutzke, G.; Fischer, B.; Mendel, R.R.; Schwarz, G. Thiocarboxylation of molybdopterin synthase provides evidence for the mechanism of dithiolene formation in metal-binding pterins. *J. Biol. Chem.* **2001**, *276*, 36268–36274. [CrossRef]
5. Rudolph, M.J.; Wuebbens, M.M.; Turque, O.; Rajagopalan, K.V.; Schindelin, H. Structural studies of molybdopterin synthase provide insights into its catalytic mechanism. *J. Biol. Chem.* **2003**, *278*, 14514–14522. [CrossRef]
6. Joshi, M.S.; Johnson, J.L.; Rajagopalan, K.V. Molybdenum cofactor biosynthesis in Escherichia coli mod and mog mutants. *J. Bacteriol.* **1996**, *178*, 4310–4312. [CrossRef] [PubMed]
7. Kuper, J.; Winking, J.; Hecht, H.J.; Mendel, R.R.; Schwarz, G. The active site of the molybdenum cofactor biosynthetic protein domain Cnx1G. *Arch. Biochem. Biophys.* **2003**, *411*, 36–46. [CrossRef]
8. Llamas, A.; Mendel, R.R.; Schwarz, N. Synthesis of adenylated molybdopterin—An essential step for molybdenum insertion. *J. Biol. Chem.* **2004**, *279*, 55241–55246. [CrossRef] [PubMed]
9. Krausze, J.; Hercher, T.W.; Zwerschke, D.; Kirk, M.L.; Blankenfeldt, W.; Mendel, R.R.; Kruse, T. The functional principle of eukaryotic molybdenum insertases. *Biochem. J.* **2018**, *475*, 1739–1753. [CrossRef]
10. Probst, C.; Yang, J.; Krausze, J.; Hercher, T.W.; Richers, C.P.; Spatzal, T.; Kc, K.; Giles, L.J.; Rees, D.C.; Mendel, R.R.; et al. Mechanism of molybdate insertion into pterin-based molybdenum cofactors. *Nat. Chem.* **2021**, *13*, 758–765. [CrossRef]
11. Palmer, T.; Vasishta, A.; Whitty, P.W.; Boxer, D.H. Isolation of Protein Fa, a Product of the Mob Locus Required for Molybdenum Cofactor Biosynthesis in Escherichia-Coli. *Eur. J. Biochem.* **1994**, *222*, 687–692. [CrossRef]
12. Neumann, M.; Mittelstadt, G.; Seduk, F.; Iobbi-Nivol, C.; Leimkuhler, S. MocA Is a Specific Cytidylyltransferase Involved in Molybdopterin Cytosine Dinucleotide Biosynthesis in Escherichia coli. *J. Biol. Chem.* **2009**, *284*, 21891–21898. [CrossRef]
13. Reschke, S.; Sigfridsson, K.G.; Kaufmann, P.; Leidel, N.; Horn, S.; Gast, K.; Schulzke, C.; Haumann, M.; Leimkuhler, S. Identification of a bis-molybdopterin intermediate in molybdenum cofactor biosynthesis in Escherichia coli. *J. Biol. Chem.* **2013**, *288*, 29736–29745. [CrossRef]
14. Eaves, D.J.; Palmer, T.; Boxer, D.H. The product of the molybdenum cofactor gene mobB of Escherichia coli is a GTP-binding protein. *Eur. J. Biochem.* **1997**, *246*, 690–697. [CrossRef] [PubMed]
15. Magalon, A.; Frixon, C.; Pommier, J.; Giordano, G.; Blasco, F. In vivo interactions between gene products involved in the final stages of molybdenum cofactor biosynthesis in Escherichia coli. *J Biol. Chem.* **2002**, *277*, 48199–48204. [CrossRef]
16. McLuskey, K.; Harrison, J.A.; Schuttelkopf, A.W.; Boxer, D.H.; Hunter, W.N. Insight into the role of Escherichia coli MobB in molybdenum cofactor biosynthesis based on the high resolution crystal structure. *J. Biol. Chem.* **2003**, *278*, 23706–23713. [CrossRef]
17. Otwinowski, Z.; Minor, W. Processing of X-ray diffraction data collected in oscillation mode. *Method Enzym.* **1997**, *276*, 307–326. [CrossRef]
18. McCoy, A.J.; Grosse-Kunstleve, R.W.; Adams, P.D.; Winn, M.D.; Storoni, L.C.; Read, R.J. Phaser crystallographic software. *J. Appl. Cryst.* **2007**, *40*, 658–674. [CrossRef] [PubMed]
19. Winn, M.D.; Ballard, C.C.; Cowtan, K.D.; Dodson, E.J.; Emsley, P.; Evans, P.R.; Keegan, R.M.; Krissinel, E.B.; Leslie, A.G.W.; McCoy, A.; et al. Overview of the CCP4 suite and current developments. *Acta Cryst. D* **2011**, *67*, 235–242. [CrossRef]
20. Murshudov, G.N.; Vagin, A.A.; Dodson, E.J. Refinement of macromolecular structures by the maximum-likelihood method. *Acta Cryst. D* **1997**, *53*, 240–255. [CrossRef] [PubMed]
21. Bailey, S. The Ccp4 Suite—Programs for Protein Crystallography. *Acta Cryst. D* **1994**, *50*, 760–763.

22. Adams, P.D.; Afonine, P.V.; Bunkoczi, G.; Chen, V.B.; Davis, I.W.; Echols, N.; Headd, J.J.; Hung, L.W.; Kapral, G.J.; Grosse-Kunstleve, R.W.; et al. PHENIX: A comprehensive Python-based system for macromolecular structure solution. *Acta Cryst. D Biol. Cryst.* **2010**, *66*, 213–221. [CrossRef] [PubMed]
23. Emsley, P.; Cowtan, K. Coot: Model-building tools for molecular graphics. *Acta Cryst. D* **2004**, *60*, 2126–2132. [CrossRef]
24. Williams, C.J.; Headd, J.J.; Moriarty, N.W.; Prisant, M.G.; Videau, L.L.; Deis, L.N.; Verma, V.; Keedy, D.A.; Hintze, B.J.; Chen, V.B.; et al. MolProbity: More and better reference data for improved all-atom structure validation. *Protein Sci.* **2018**, *27*, 293–315. [CrossRef] [PubMed]
25. Concepcion, J.; Witte, K.; Wartchow, C.; Choo, S.; Yao, D.F.; Persson, H.; Wei, J.; Li, P.; Heidecker, B.; Ma, W.L.; et al. Label-Free Detection of Biomolecular Interactions Using BioLayer Interferometry for Kinetic Characterization. *Comb. Chem. High. T Scr.* **2009**, *12*, 791–800. [CrossRef]
26. Krissinel, E.; Henrick, K. Inference of macromolecular assemblies from crystalline state. *J. Mol. Biol.* **2007**, *372*, 774–797. [CrossRef] [PubMed]
27. Walker, J.E.; Saraste, M.; Runswick, M.J.; Gay, N.J. Distantly Related Sequences in the Alpha-Subunits and Beta-Subunits of Atp Synthase, Myosin, Kinases and Other Atp-Requiring Enzymes and a Common Nucleotide Binding Fold. *EMBO J.* **1982**, *1*, 945–951. [CrossRef]
28. DeLano, W.L.; Lam, J.W. PyMOL: A communications tool for computational models. *Abstr. Pap. Am. Chem. S* **2005**, *230*, U1371–U1372.

Article

The Depth-Dependent Mechanical Behavior of Anisotropic Native and Cross-Linked HheG Enzyme Crystals

Marta Kubiak [1,*], Marcel Staar [2], Ingo Kampen [1], Anett Schallmey [2] and Carsten Schilde [1]

[1] Institute for Particle Technology, Technische Universität Braunschweig, Volkmaroder Str. 5, 38104 Braunschweig, Germany; i.kampen@tu-braunschweig.de (I.K.); c.schilde@tu-braunschweig.de (C.S.)
[2] Institute of for Biochemistry, Biotechnology and Bioinformatics, Technische Universität Braunschweig, Spielmannstr. 7, 38106 Braunschweig, Germany; m.staar@tu-braunschweig.de (M.S.); a.schallmey@tu-braunschweig.de (A.S.)
* Correspondence: marta.kubiak@tu-braunschweig.de; Tel.: +49(0)-53139165533

Abstract: Enzymes are able to catalyze various specific reactions under mild conditions and can, therefore, be applied in industrial processes. To ensure process profitability, the enzymes must be reusable while ensuring their enzymatic activity. To improve the processability and immobilization of the biocatalyst, the enzymes can be, e.g., crystallized, and the resulting crystals can be cross-linked. These mechanically stable and catalytically active particles are called CLECs (cross-linked enzyme crystals). In this study, the influence of cross-linking on the mechanical and catalytic properties of the halohydrin dehalogenase (HheG) crystals was investigated using the nanoindentation technique. Considering the viscoelastic behavior of protein crystals, a mechanical investigation was performed at different indentation rates. In addition to the hardness, for the first time, depth-dependent fractions of elastic and plastic deformation energies were determined for enzyme crystals. The results showed that the hardness of HheG enzyme crystals are indentation-rate-insensitive and decrease with increases in penetration depth. Our investigation of the fraction of plastic deformation energy indicated anisotropic crystal behavior and higher irreversible deformation for prismatic crystal faces. Due to cross-linking, the fraction of elastic energy of anisotropic crystal faces increased from 8% for basal faces to 68% for prismatic crystal faces. This study demonstrates that mechanically enhanced CLECs have good catalytic activity and are, therefore, suitable for industrial use.

Keywords: halohydrin dehalogenase (HheG); enzyme; immobilization; cross-linked enzyme crystal (CLEC); micromechanics; nanoindentation; catalytic activity

Citation: Kubiak, M.; Staar, M.; Kampen, I.; Schallmey, A.; Schilde, C. The Depth-Dependent Mechanical Behavior of Anisotropic Native and Cross-Linked HheG Enzyme Crystals. *Crystals* **2021**, *11*, 718. https://doi.org/10.3390/cryst11070718

Academic Editors: Kyeong Kyu Kim and T. Doohun Kim

Received: 18 May 2021
Accepted: 19 June 2021
Published: 22 June 2021

Publisher's Note: MDPI stays neutral with regard to jurisdictional claims in published maps and institutional affiliations.

1. Introduction

Halohydrin dehalogenases (HHDHs) (E.C. 4.5.1.-) are bacterial lyases that belong to the superfamily of short-chain dehydrogenases and reductases [1]. Apart from the degradation of toxic halogenated compounds, the industrial relevance of these lyases is based on their epoxide ring opening activities with a wide range of different nucleophiles enabling the formation of novel C–C, C–N, C–O, and C–S bonds [2]. Here, halohydrin dehalogenase HheG from *Ilumatobacter coccineus* is of special interest due to its ability to accept cyclic and other sterically demanding epoxide substrates such as vicinally di-substituted epoxides, further broadening the accessible product range of β-substituted alcohols [3–5]. Moreover, HheG is able to catalyze the α-regioselective ring-opening of different styrene oxide derivatives, with cyanate yielding the corresponding oxazolidinones [6]. The resulting products represent important compounds for the synthesis of fungicides and antibiotics [7,8]. Moreover, oxazolidinones are used as chiral auxiliaries in chemical synthesis [9]. Due to the wide range of its use, this enzyme is of special interest for future industrial applications.

To reach their full industrial potential, enzymes must meet various requirements for industrial applications, such as shear force resistance in a bioreactor and reusability. For this reason, soluble enzymes can be immobilized on a carrier or cross-linked as protein particles, either as crystals (CLECs) or aggregates (CLEAs) [10]. The mechanical behavior of protein crystals or aggregates depends on the structure of the individual protein molecules, the packing density, and the conformation within the three-dimensional structure of the protein particles [11] and must, therefore, be investigated individually. Since the mechanical stability of particulate biocatalysts is a fundamental property required for industrial applications, knowledge of the mechanical properties of these biocatalysts is of great interest. Previous studies have reported on the micromechanics of native protein crystals [12–17], as well as their fragility and sensitivity to environmental changes [18,19]. In recent years, there has been increasing interest in enhancing mechanical stability by cross-linking enzyme crystals [20,21]. This immobilization method offers stabilization of the crystalline enzyme structure while maintaining catalytic activity [20,22]. This technology is complementary to protein engineering methods that aim at boosting the inherent stability of an enzyme [23]. One of the most commonly used linkers for the immobilization of enzyme crystals described in the current literature is glutaraldehyde [24,25] because of its low cost, ease of handling, and high efficacy [26]. Morozov et al. were the first to use a resonance technique to measure parent (native) and cross-linked triclinic lysozyme crystals, as well as the Young's modulus of the parent protein crystals in a range of 290−1400 MPa. The authors also reported no significant influence from the cross-linking of protein crystals on this value range [27]. In contrast, Lee et al. investigated the breakage probability of cross-linked yeast alcohol dehydrogenase (YADH) crystals through shearing with a rotating disc device. The authors observed no breakage of rod-shaped CLECs but did record the breakage of hexagonal CLECs at energy dissipation rates above 0.1 $MW \cdot kg^{-1}$. The authors also studied non-cross-linked YADH crystals and noted a significant shift of the particle size distribution toward smaller particle sizes compared to those of cross-linked YADH [28]. Kubiak et al. measured the micromechanical properties of different CLECs (distinct PGA variants, Lysozyme, and HheG) using atomic force microscopy-based (AFM) nanoindentation. The determined differences in the mechanical behaviors of distinct protein crystals were correlated with the different crystal structures and properties, such as water content and the number of lysine residues able to be cross-linked [29,30]. Furthermore, the authors identified the anisotropic mechanical behaviors of the different HheG crystal faces, which were caused partly by the crystal structure and partly by the cross-linking time [31].

The cross-linking of enzyme crystals has an influence on the crystal's mechanical properties and catalytic activity. An increase in cross-linking time leads to improved mechanical stability, but excessive cross-linking may cause protein precipitation and a loss in activity [32]. To use a biocatalyst successfully in an industrial process, both properties, mechanical stability and catalytic activity, must be present. In the context of this publication, the mechanical and catalytic properties of native and cross-linked enzyme crystals based on the HheG wildtype were examined, and the influence of cross-linking on the crystal properties was determined. For this purpose, a Nanoindenter was used as a tool to relate structural changes to variations in material properties due to this device's ability to accurately measure micromechanical properties at various length scales [33]. This study compares the micromechanical properties of native and cross-linked protein crystals, which were examined under the same conditions using the nanoindentation technique. To the best of our knowledge, we are the first to evaluate the depth-dependent elastic and plastic functionality of indentation. Moreover, the catalytic properties of the designed CLECs are examined to fully characterize an industrially relevant HheG enzyme.

2. Materials and Methods

2.1. Crystallization and Cross-Linking

Halohydrin dehalogenase HheG from *Ilumatobacter coccineus* (HheG) was heterologously produced in *Escherichia coli* BL21(DE3) and purified to homogeneity as described previously [4]. Crystallization of the HheG wildtype was performed using the sitting drop method. A 20 µL droplet composed of protein stock solution (32 mg/mL) and precipitation solution (PEG 4000 (10% (w/v) in a HEPES buffer (10 mM, pH 7.3)) was placed on the cover slide and equilibrated against the reservoir solution (500 μL) at 5 °C. After 24 to 72 h, hexagonal crystals with an average size of 70 μm, as seen in Figure 1, were observed under an optical microscope and prepared for cross-linking using a gentle vapor diffusion method, based on the work of Lusty et al. [34]. For cross-linking, crystallization microplates equipped with microbridges in single wells were used. Crystals grown on a cover slide were washed on ice three times using a precipitation solution to remove the mother liquor and non-crystalized protein. Our experience has shown that cooling the crystals on ice is necessary to avoid crystal breakage caused by osmotic shock during the washing step [35]. After crystal washing, 10 μL of the fresh precipitation solution was added to the crystals. Then, 25 μL of a 25% aqueous glutaraldehyde solution (glutaraldehyde diluted in precipitation solution, pH~3 adjusted with HCl (1M)) was added to the well of the microbridge. According to Monsan et al., in an acidic pH solution, glutaraldehyde is present as a free aldehyde [36]. Finally, the cover slide containing the washed crystals was sealed over the well of a microplate. The microplate was then immediately turned over to allow the crystals to cross-link for approximately 24 h at 5 °C as a sitting droplet.

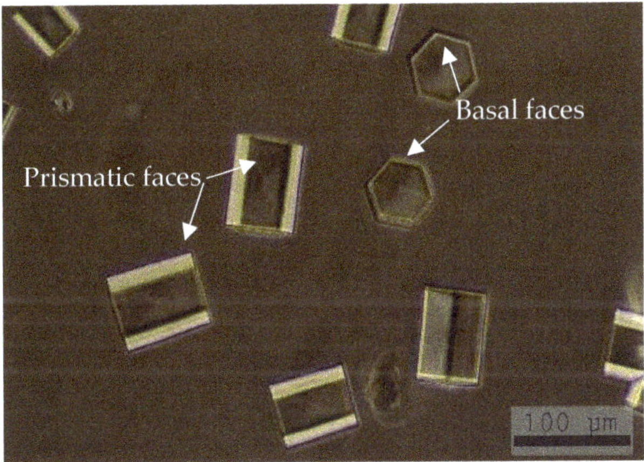

Figure 1. Prismatic (rectangle) and basal (hexagon) crystal faces of the HheG wildtype crystals. Reprinted with permission from [29]. Copyright (2018) American Chemical Society.

2.2. *Bioconversion of Cyclohexene Oxide Using HheG CLECs*

Conversion of the cyclohexene oxide using HheG CLECs was carried out at a 1 mL scale, and the reactions were analyzed by gas chromatography (GC) [37]. Reactions were performed at 22 °C and 900 rpm in a Tris·SO_4 buffer (50 mM, pH 7.0) containing cyclohexene oxide (20 mM) and sodium azide (40 mM) as a nucleophile. According to the reaction, we applied all HheG CLECs obtained from one crystallization and cross-linking batch, starting from the 320 μg soluble HheG. For comparison, a reaction using 100 μg of soluble HheG was performed as a positive control. However, the negative control reactions performed to investigate the chemical background did not contain the enzyme. After 4 and 24 h, samples of each 400 μL solution were taken and extracted with an equal volume of *tert* butyl methyl ether (TBME) containing 0.1% dodecane as an internal standard. After drying the organic extracts over $MgSO_4$, the samples were analyzed by GC.

2.3. *Mechanical Analysis on Single Protein Crystals*

A Hysitron TriboIndenter Ti900 equipped with a Berkovich tip was used for the nanoindentation tests of native and cross-linked enzyme crystals. To avoid mechanical interventions, the crystals were indented on the cover slide upon which they grew. Since the crystals were grown on a siliconized slide in a sitting droplet, they adhered to the surface and were 'immobilized', which is needed for reliable indentation measurements.

The cover slide was then mounted on a heating–cooling stage, which had a temperature of 5 °C. Because the indentation was performed in a viscous liquid (a PEG solution for native crystals), a preload of 5 µN was set for surface detection. Measurements at lower forces were not possible due to frequently according errors in surface detection, which resulted in a loss of contact between the Berkovich tip and the crystal surface during the measurement. To quantify the indentation size effect, indentation depths from 100 to 900 nm were set as a pattern (3 × 3 with gaps of 10 µm) on each single crystal face. The maximum penetration depth was set as 900 nm to avoid breakage of the native crystals. For this reason, the displacement-controlled mode was applied for the mechanical examination. Additionally, to investigate the influence of the indentation rate on the mechanical behavior of viscoelastic enzyme crystals, different displacement rates were applied for the measurements. It was not possible to use very fast indentation rates at small penetration depths (100–300 nm) because of machine compliance, which required an indentation duration time of 0.2 s or less. For this reason, the measurements were performed with either a constant displacement rate of 180 nm/s or a constant segment time of 0.55 s, meaning that each measurement was performed with an increased displacement rate, starting with 180 nm/s at a penetration depth of 100 nm up to 1636 nm/s at 900 nm, as shown in Figure 2.

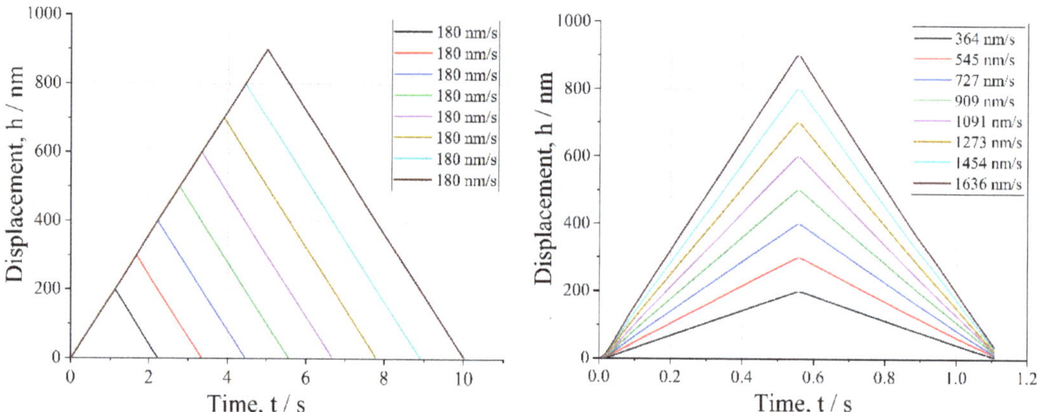

Figure 2. Displacement functions of applied modes: constant displacement rate (**left**) and constant segment time (**right**).

For each indentation mode, nine measurements were taken on each of the approximately 18 crystals, distinguishing between basal and prismatic crystal faces. Based on the recorded force–displacement curves, the indentation hardness was calculated according to the Oliver and Pharr theory [38]. Using the example of hardness, Figure 3 shows that this number of indentations per penetration depth ensures sufficient statistical certainty—i.e., both the mean value and the standard deviation assume a constant value. Since the standard deviation is a measure of the distribution of values, this is the crucial point at which distributions must be compared with each other.

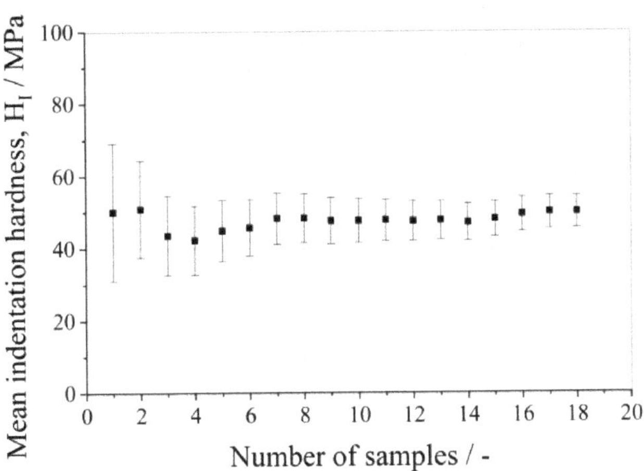

Figure 3. Influence of the number of samples on the confidence interval around the mean value of indentation hardness for basal faces at a penetration depth of 200 nm (confidence interval of 95%).

Additionally, elastic and plastic indentation work were calculated according to Formula (1) [39]:

$$E_{elastic} = \int_0^{h_{max}} F_{unload}(h)dhE_{plastic} = \int_0^{h_{max}} F_{load}(h) - F_{unload}dhE_{total} \quad (1)$$

where h_{max} is the maximum displacement during indentation, F_{load} and F_{unload} are the indentation forces during loading and unloading of the sample, elastic deformation is defined as the area under the unloading curve, and plastic deformation is defined as the area enclosed between loading and unloading curves, as shown in Figure 4.

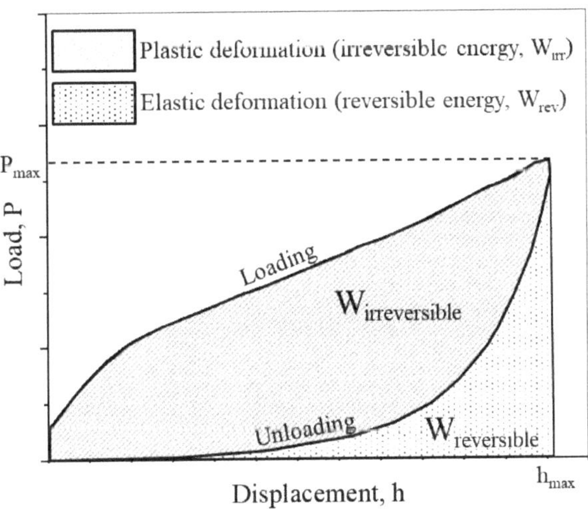

Figure 4. Determination of plastic and elastic work in the instrumented indentation testing.

All results are presented as a boxplot, which is explained in Figure 5.

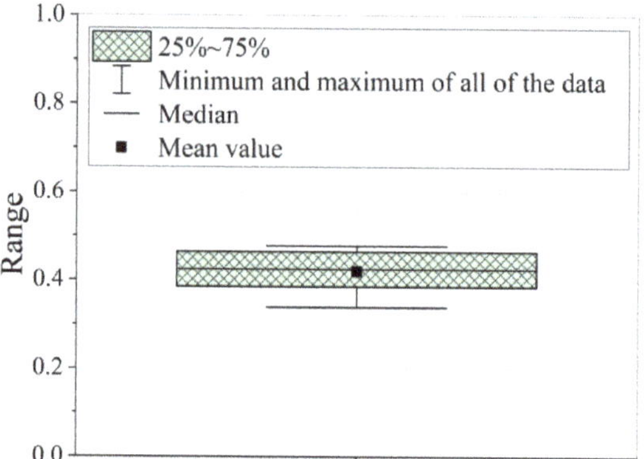

Figure 5. Graphical description of the variation between samples using the boxplot method.

The measurements at a penetration depth of 100 nm were discarded because of the temperature drift of the indenter tip during the first measurement, which was caused by transferring the tip from room temperature to the droplet with a temperature of 5 °C.

3. Results and Discussion

3.1. Influence of Displacement Rate on Depth-Dependent Mechanical Response

The time-dependent properties of protein crystals, such as their viscoelastic behaviors and creep, have been noted in many previous studies [12,27,40]. Although creep seems to have no significant impact on mechanical responses at small scales [29], it could affect measurements at large penetration depths. To establish the guidelines for further measurements, we first investigated and analyzed the influence of the displacement rate on mechanical properties in detail. Figure 6 presents the indentation hardness of native and cross-linked prismatic crystal faces measured at different penetration depths and displacement rates. Notably, under a constant segment time of 0.55 s, the displacement rate increased, as shown in Figure 2.

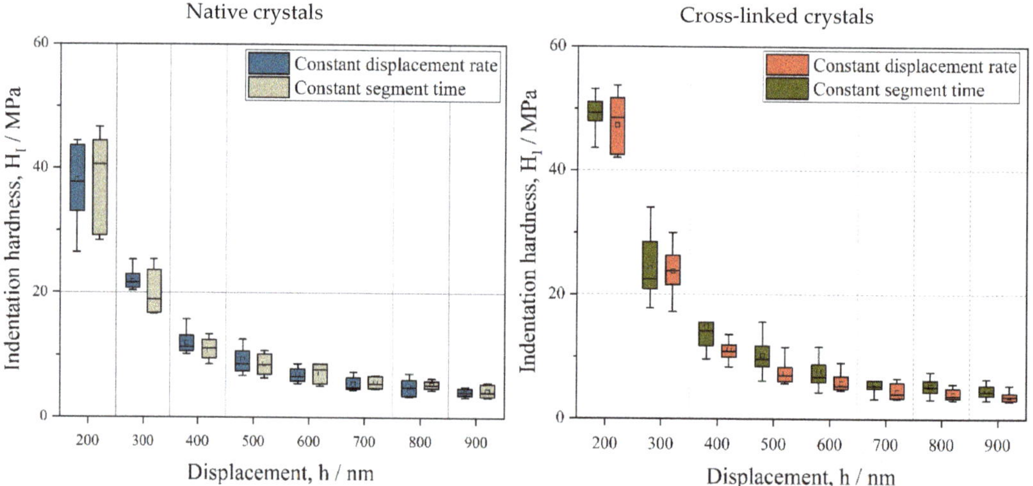

Figure 6. Indentation hardness of the prismatic faces of native (**left**) and cross-linked (**right**) crystals as a function of the depth and indentation displacement rate. The indentation rate of the constant-displacement-rate mode was 180 nm/s over the whole penetration range, while the indentation rate of the constant-segment-time mode increased from 364 nm/s at a penetration depth of 200 nm to 1636 nm/s at a penetration depth of 900 nm.

Here, we can observe three effects: (1) the tendency of hardness to decrease with an increase in penetration depth, (2) the higher hardness of cross-linked HheG crystals only at small penetration depths, and (3) the insensitivity of indentation hardness to the indentation rate. For both measurement modes, the indentation hardness decreases from ca. 39 MPa (mean value) for native crystals and 48 MPa for cross-linked crystals at a penetration depth of 200 nm to 4 MPa at high displacements of 900 nm. This so-called indentation size effect (ISE) has been widely examined and modeled using, e.g., the concepts of statistically stored dislocations (SSTs) and geometrically necessary dislocations (GNDs) [41]. SSTs are homogenously distributed in materials and dependent upon the material and processing conditions. Geometrically necessary dislocations (GNDs) are dislocations that must be present near the indentation to accommodate the volume of material displaced by the indenter at the surface [41]. The basic principle underlying this model is that the GNDs exist in addition to the usual statistically stored dislocations (SSDs) produced under uniform strain, giving rise to an extra hardening component that becomes larger as the contact impression decreases in size. The dislocation density is generally defined as the quotient of the dislocation length and volume in which the dislocations are stored. Thus, the geometrically necessary dislocation density (ρ_G) for an indentation with a conical indenter is defined as follows:

$$\rho_G = \left(3\tan^2\theta\right)/(2bh) \tag{2}$$

where b is the Burger's vector, h is the penetration depth, and θ is the angle between the indenter and the deformed surface. According to formula 2, the hardness increases at small depths because the geometrically necessary component of the dislocation density is inversely proportional to the depth and increases when the contact is minimal [41].

Regarding the magnitude of hardness, there are no comparable data showing depth-dependent hardness of protein crystals, especially HheG crystals. Nevertheless, Table 1 summarizes some of the hardness results of native and cross-linked lysozyme and HheG crystals. According to the lysozyme results, the hardness strongly depends on the crystal structure. For example, the Vickers microhardness of the (010) plane of triclinic (tri-) HEWL crystals is over 4 times greater than the hardness of the (010) plane of orthogonal (O-) HEWL crystals [42,43]. In the cited studies, the hardness was measured in deep regions of the crystals (about 700 nm) via nanoindenter measurements [12]. For this reason, the HheG hardness at a penetration depth of 900 nm should be compared with the values in the literature. The hardness in the deep regions of native HheG crystals is about 4 MPa, which is comparable to the hardness of the (110) plane of O-HEWL crystals (6 MPa) [43]. According to Suzuki et al., hardness is an intrinsic property related to the habit planes and, therefore, is also related to anisotropy due to molecular packing and intracrystalline water [42]. The slightly lower hardness of HheG crystals can be explained by the high solvent content of the HheG crystals (65%) compared to the O-HEWL crystals (43%) [44].

Table 1. Summary of the hardness results for the native and cross-linked lysozyme and HheG crystals.

Enzyme Crystal	Specifics	Measuring Method	Hardness/MPa	Literature
Lysozyme	Native, wet triclinic crystals	Micro-Vickers hardness	25–28	[42]
Lysozyme	Native, wet orthorombic crystals	Micro-Vickers hardness	6–10	[43]
Lysozyme	Native, wet tetragonal crystals	Nanoindenter	15	[12]
Lyzozyme	Cross-linked tetragonal crystals in liquid	AFM Nanoindentation	11	[29]
HheG	Cross-linked prismatic crystal faces	AFM Nanoindentation	8	[31]

There is good agreement between the hardness of the cross-linked HheG crystals studied with the help of nanoindentation and that studied using AFM. (4 vs. 8 MPa).

However, the hardness was measured in the submicrometer range using a nanoindenter, and the results of AFM were obtained at a nanometer scale. By using the spherical tip instead of the Berkovich indenter, the indentation size effect, caused in part by the geometry of the tip, can be reduced. According to Swadener et al., the spherical indenter uniquely shows no depth dependence on hardness but does show dependence on the radius of the sphere [45]. This well-known effect provides an increase in hardness with a decrease in the tip radius [46]. This phenomenon explains why the hardness measured with the Berkovich tip (with a tip radius of ca. 50 nm) at low penetration depths of 200 nm was five times higher than the AFM hardness (with a tip radius of ca. 150 nm).

The last factor is the influence of cross-linking on the hardness of HheG crystals. Cross-linking increased the hardness by about 20% near the crystal surface at a penetration depth of 200 nm. In deeper regions (e.g., from ca. 400 nm), there was no significant difference in the hardness values. To provide a better understanding, this phenomenon is explained in Section 3.3. based on the many results presented in later parts of this manuscript.

The most surprising aspect of the data is the insensitivity to the displacement rate. While the present results are consistent with the findings of Raut et al., who examined organic crystals using the nanoindentation technique with a quasi-static load, they are at odds with the intrinsic time-dependent behavior of the crystals [47]. Raut et al. examined the strain-rate sensitivity of anorganic crystals. The authors reported that a lack of strain-rate sensitivity in crystals correlates with very low intermolecular interactions, such as van der Waals and hydrogen bonds, which can break easily such that plastic deformation occurs through the shearing of specific crystallographic planes [47]. An atomistic study of a protein dimer was previously published by Buehler et al. [48]. Although, in mechanical terms, protein crystals behave somewhat differently from individual proteins in their overall construction, deformation can be facilitated by the motion of individual molecules. Thus, interactions at the smallest scale, such as those within a protein dimer, can be considered in this context. The aforementioned study demonstrated the dependence of force–strain curves on pulling speed only above a velocity of a 1 m/s [48]. Moreover, the authors explained the unfolding process in terms of its dependence on the pulling speed. While at large velocities, unfolding occurs through the sequential breaking of hydrogen bonds one by one, at lower velocities (under 0.161 m/s), all hydrogen bonds are stretched equally and fail simultaneously [48]. However, although there was no detectable correlation between the applied displacement rate and the indentation hardness of enzyme crystals in the investigated range, different behavior could potentially be measured at significantly higher or lower displacement rates. Moreover, the penetration rate may have an influence on adhesion. During the data analysis, we noted that the tendency towards adhesion was higher at lower indentation rates. Figure 7 provides an example comparison of the representative force displacement curves at slow (180 nm/s) and fast (1273 nm/s) displacement rates.

It can be clearly seen that the force–displacement curve of the constant displacement rate measurement mode shows adhesion and a slight deviation from the almost smooth loading curve of the constant segment time mode. The latter result indicates that the tendency for dislocations to occur increases with an increase in contact time between the indenter tip and the specimen. Considering that the rate of displacement has no effect on the mechanical properties and that low displacement rates may cause adhesion, only the results of the constant segment time mode are presented in the following portion of the manuscript.

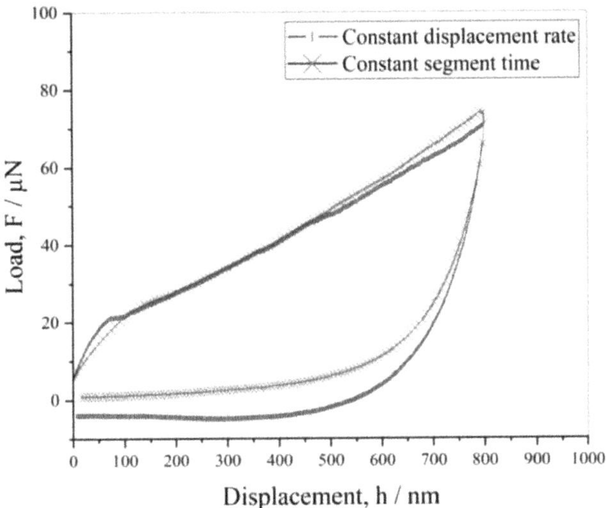

Figure 7. Comparison of representative force–displacement curves of prismatic crystal faces at slow (180 nm/s, constant displacement rate) and fast (1273 nm/s, constant segment time) displacement rates at a penetration depth of 800 nm.

Due to the depth-dependence of mechanical measurements, in subsequent graphs, the fraction of elastic deformation in the indentation work is shown instead of the elastic modulus. Moreover, the indentation hardness, which is correlated with the elastic modulus [49], is replaced by the fraction of plastic deformation of an indentation. Hence, all subsequent results are presented using the elastic and plastic fractions of deformation energy.

3.2. Anisotropic Behavior of Native HheG Crystals

To determine the changes in mechanical behavior due to cross-linking, we first investigated the mechanical behaviors of both faces of the native crystals. Figure 8 compares the energy fraction of the irreversible deformation of the prismatic and basal crystal faces. As expected, an increased fraction of plastic deformation energy with an increase in displacement is observable in the data for both crystal faces. The fraction of irreversible plastic deformation energy of the basal and prismatic faces starts at 36% and 63% and then increases to 62% and 78%, respectively. Interestingly, compared with the prismatic face, the basal face exhibits about 43% lower plastic deformation energy at relatively small penetration depths of 200 nm. This difference becomes smaller with an increase in penetration depth—ending at only 15%, at a depth of 900 nm.

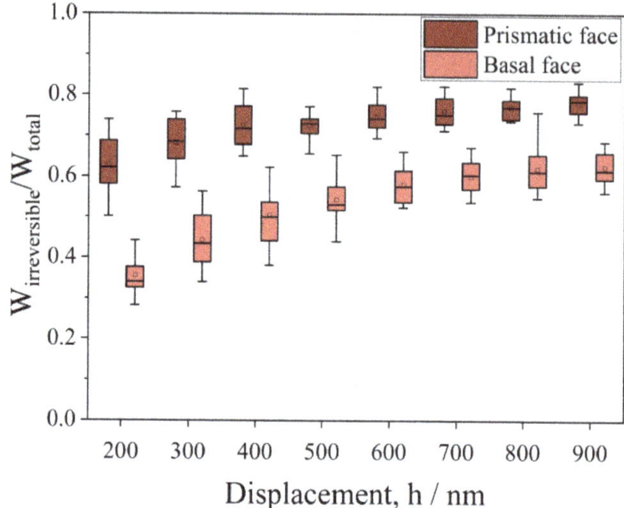

Figure 8. Fraction of irreversible deformation on the prismatic and basal crystal faces at penetration depths from 200 to 900 nm.

There are two fundamental reasons for the observed anisotropy. Anisotropy can be related to either molecular packing or the intermolecular forces between the proteins, which can be different in each direction. Many publications have described the anisotropy of molecular crystals in the context of hydrogen bonds [50–53]. In the following section, the individual factors of this phenomenon will be discussed in more detail. Previously, Buehler studied an atomistic model of protein crystals and reported that plastic deformation is caused by chain unfolding [54], which indicates that permanent deformation occurs at a molecular scale. However, the experimental results presented in our manuscript contradict the atomistic modeling results provided by Buehler, who described the uniaxial strain of a perfect protein crystal of a small protein α-conotoxin PnIB from *conus pennaceus* in various directions [54]. Based on the simulated strain–stress curves, a comparable elastic modulus was observed for small deformations in all crystallographic orientations along with very different mechanical behaviors for large deformations [54]. As illustrated in Figure 8, the anisotropic faces show a greater difference in mechanical deformation under small than large deformations. This result is the opposite of the results in our study. Because no experimental data on the mechanical behavior of the crystals were provided in the aforementioned study, it could not be clarified whether these differences were due to the different behaviors of the distinct protein crystals or due to the inaccurate force fields used to model the crystal's deformations. According to the author, future studies could be focused on using more accurate force fields that offer the possibility of bond breaking and formation for modeling deformation [54].

Hence, a mechanism for anisotropic plastic deformation that considers the sliding of crystallographic planes should be additionally considered and discussed. For this purpose, the mechanical behavior is first analyzed on the basis of the force–displacement curves. Figure 9 provides example force–displacement curves of nanoindentation on native prismatic and basal crystal faces.

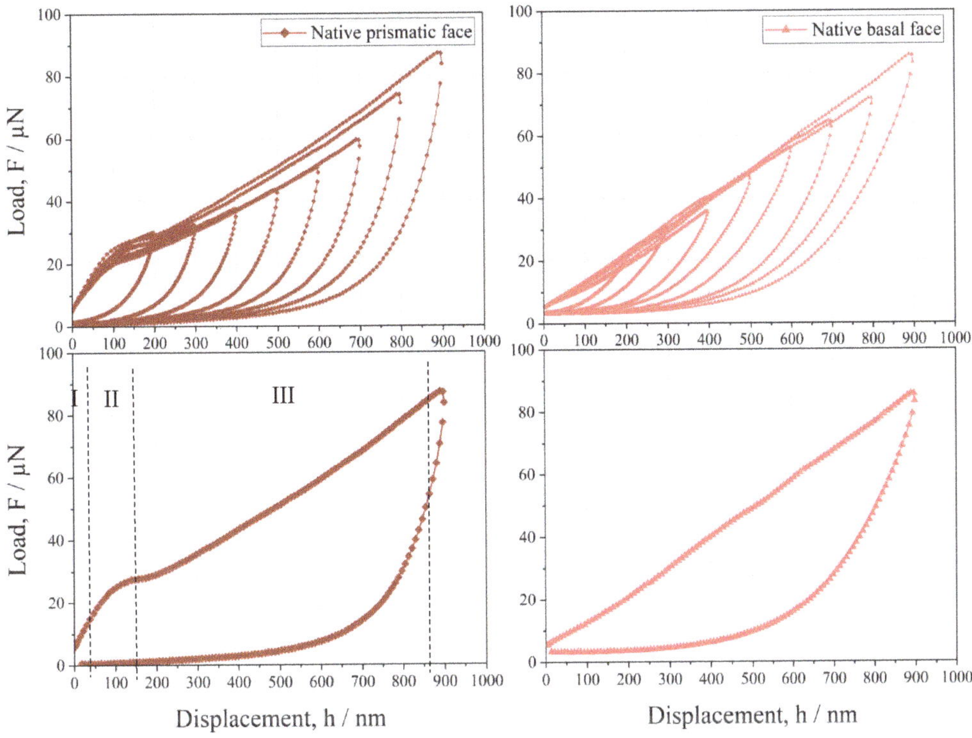

Figure 9. Representative force–displacement curves of native prismatic (**left**) and basal (**right**) crystal faces. Bottom: single curves at a penetration depth of 900 nm are presented for better resolution of the mechanical response.

Since the original studies by G. I. Taylor [55,56], numerous attempts have been made to explain the work-hardening phenomena in terms of dislocation mechanisms. Assuming that the deformation behaviors of organic and inorganic crystals are comparable [57], we analyzed and interpreted the deformation behavior of protein crystals using the typical stress–strain curves of an FCC metal crystal. In the case of the prismatic face (see Figure 9), nearly complete elastic behavior up to penetration depth of ca. 100 nm was measured (region I). In this case, the critical resolved shear stress was exceeded, allowing us to observe the yield strength. According to Taylor's theory, flattening of the loading curve indicates that only one slip system is active in a given region and that dislocations migrate to a favorable oriented slip plane (region II). At a penetration depth of ca. 150–200 nm, as seen in Figure 9, a further increase in force is required for indentation (region III). The higher the penetration depth (and hence, the displacement rate) is in the constant-segment time mode, the higher the required force will be. Based on the theory of Taylor, this is a result of multiple gliding, where several slip systems are deformed simultaneously [55,56]. This phenomenon can be observed because the slip planes have no time for reorientation, as previously unfavorably located slip planes can now be displaced into more favorable positions. As a result, mutual hindrance of the dislocation movements leads to increased force. Here the stress decreases, unlike in region I, where the slope of the loading curve is much higher. In contrast, the loading curves of the basal face do not indicate any significant hindrance of slip movements. Moreover a much higher elastic fraction can be seen in Figure 9. These results can be explained by the fact that the anisotropic properties correlate with the crystal packing. Previous studies reported the dependence of hardness on the indented crystal plane [13,15,17,43,57]. These past studies showed a relationship between plastic deformation, dislocation multiplication, and motion in inducing a slip.

Unfortunately, there is no detailed discussion on the correlation between the hardness and dislocation motion in each slip system within a protein crystal. However, the crystal packing shown in Figure 10 indicates that anisotropic deformation may be related to slip planes with different arrangements.

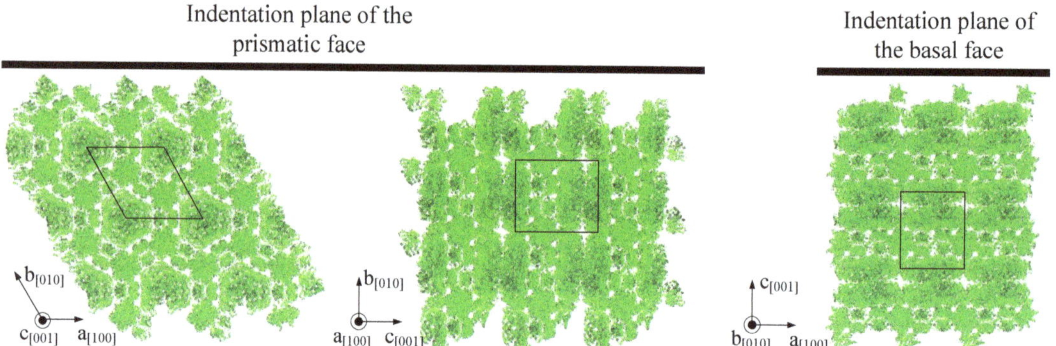

Figure 10. Molecular packing of HheG crystals with marked boundaries of the unit cell. The black line shows a trace of the prismatic (**left**) and basal (**right**) face on which indentation was performed. The indices a, b, and c refer to the three edges of the unit cell and the corresponding IJK indicies.

Xia et al. examined FCC metal using a nanoindenter equipped with a spherical diamond tip and categorized the plastic flow for each slip plane by the inclination angle of that plane to the surface [58]. The authors reported that the flow direction transformed from forward motion to sideways motion at a transition angle of 55° to 58° [58]. The graph shows that the indentation plane of the prismatic face is aligned parallel to the slip plane [010], indicating a sideways and hampered flow. In the case of a basal face, the indenter load is aligned orthogonally to the slip plane [001], where a forward flow (and easy gliding) can be expected.

HheG enzymes are large homotetramer molecules with a molecular weight of 299.37 kDa. These enzymes crystallize in a trigonal crystal system with the space group $P3_121$. As a result, the crystals exhibited significant solvent content of 65% [4]. Nanev reported that intra-crystalline water works as a "lattice glue" and helps to hold the crystal structure together [59]. This effect is based on the dynamic hydrogen bond chains that form between some intra-crystalline water molecules, which are likely to become elongated to connect the amino-acid residues of adjacent protein molecules in the crystal lattice [59]. This effect is expected to be observed within very small channels and gaps where the molecules are at an appropriate distance to each other to be connected. These results show that there are many possible reasons for anisotropy and that the directional differences in mechanical behavior are most likely due to the interplay between the structure and the resulting crystal contacts.

3.3. Mechanical and Catalytic Properties of Cross-Linked HheG Crystals

Figure 11 summarizes all the results for the fraction of the reversible (elastic) indentation energy of native and cross-linked HheG crystals. First, it can be observed that the elastic energy decreases with an increase in penetration depth for both kinds of crystals. Second, it is evident from the figure that both faces of the cross-linked crystals exhibit higher elastic deformation energy than the native crystals. Indeed, the elastic energy of the basal faces shows the highest fraction and ranges from 72% to 42%. Next highest are the prismatic faces, ranging from 66% to 37%. The native basal faces of the crystals have similar properties, where the fraction of elastic energy decreases from 64% to 39%. The native prismatic surfaces of the crystals exhibit the lowest fraction of elastic energy, which decreases in the investigated range from 37% to 22%. The third and most surprising aspect of this study is the order of magnitude by which the native prismatic faces were reinforced. Due to cross-linking, the elastic energy of prismatic faces increased by about 78% at a penetration depth of 200 nm to 68% at high penetration depths. Moreover, cross-linking strengthened the crystal lattice against fractures. This was determined after none of the examined crystals presented cracks or fractures.

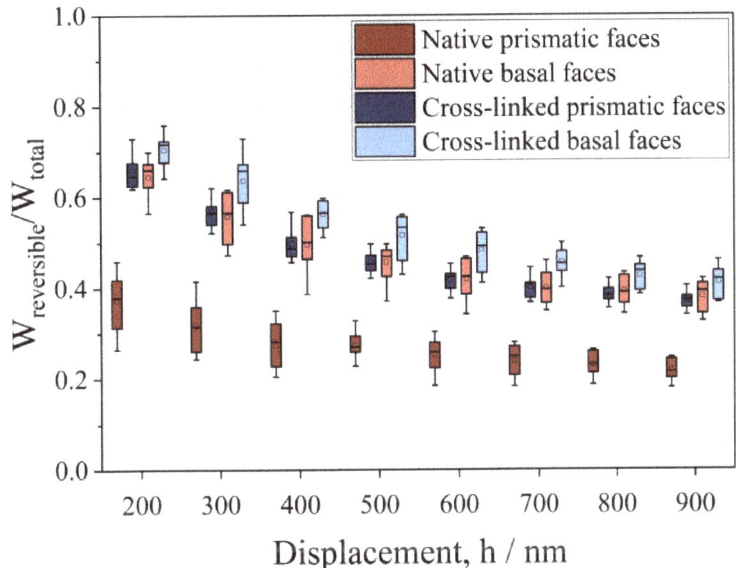

Figure 11. Fraction of the reversible indentation work of native and cross-linked enzyme crystals as a function of penetration depth.

As mentioned in previous literature, due to cross-linking, strong covalent bonds become inserted into the crystal lattice [21,60,61]. Cross-linking bonds can occur intramolecularly (between groups in the same protein molecule) or intermolecularly (between different protein molecules) [62], where the latter are necessary to maintain the crystal packing in environments different from those of the crystallization liquor [23]. Kubiak et al. developed a mathematical model for calculating theoretical crystal strength based on the potential cross-linking bonds between three residual pairs: the ε-amines of lysine residues (Lys–Lys), two neighboring arginine residues (Arg–Arg), and arginine and lysine residues (Arg–Lys). Using the MATLAB toolbox, the authors analyzed the distance and direction of possible cross-linking bonds. From this analysis, the dominant crystal faces were determined, showing that higher number of perpendicular bond fractions yields a crystal face with greater strength. A summary of the direction-dependent crystal-strength results for the considered residual pairs is provided in Table 2 [31].

Table 2. Direction-dependent theoretical crystal strength due to the anisotropic cross-linking bonds between three residual pairs: arginine–arginine, arginine lysine, and lysine–lysine. Surf 1 is the basal face, and Surf 2–3 are the three prismatic faces. Reprinted with permission from [31]. Copyright (2019) American Chemical Society.

Crystal Face	Arg–Arg	Arg–Lys	Lys–Lys	Sum
Surf 1	271.77	142	145.4	559.17
Surf 2	231.63	137.81	107.04	476.48
Surf 3	230.52	140.01	105.47	476.00
Surf 4	230.49	141.84	107.97	480.30

Assuming that all the bonds in deep crystal regions contribute to stabilizing the crystal structure, the basal face (Surf 1~560) has about a 15% higher fraction of anisotropic cross-linking bonds than the prismatic faces (Surf 2–Surf 4~477) [31]. This would result in a basal crystal face with higher strength, which agrees with our experimental results for cross-linked crystals. Mechanical analysis of the cross-linked anisotropic faces showed that the elastic fraction of deformation energy was ca. 15% higher for the basal faces than the prismatic faces.

We also identified most of these bonds near large and small channels within the crystal, as shown in Figure 12 [31].

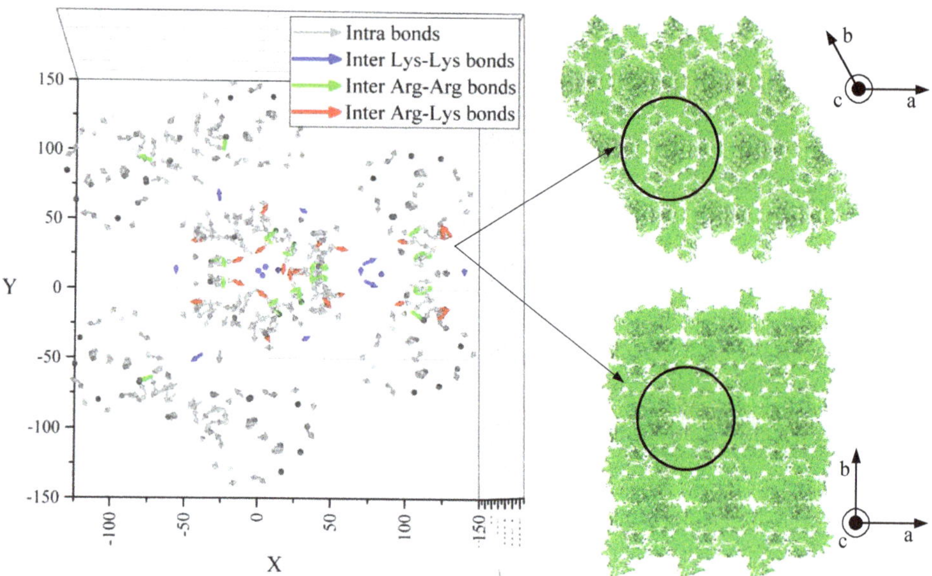

Figure 12. Graphical illustration of the intra- and intermolecular bonds with a maximum distance of 10 Å within the crystal packing composed of several protein molecules (**left**) and their possible occurrence within a supercell (**right**). Reprinted with permission from [31]. Copyright (2019) American Chemical Society.

Due to the strong covalent bonds, it seems possible to hold the crystallographic slips together such that the cross-linked prismatic faces exhibit much stronger elastic recovery than the native prismatic faces. This mechanical response can be clearly seen on the force displacement curves shown in Figure 13.

Compared to the force–displacement curves of native crystals, the loading and unloading curve of the cross-linked crystals shows a significantly higher elasticity fraction, higher forces required to create an indentation, and no flattening curve section. The shape of the stress curve of the cross-linked crystals is similar to that of polycrystalline materials, in which multi-gliding occurs immediately after elastic deformation. It can thus be suggested that cross-linking bonds, which lead to a decrease in the effective "pore" size of the channels [22], enclose the mobile water within a crystal lattice, which is why the crystal behaves like a damper against plastic deformation. Another possibility is that cross-linking bridges formed randomly within the crystal can transmit forces in all directions. In this way, strong covalent bonds contribute to a reduction in plastic deformation.

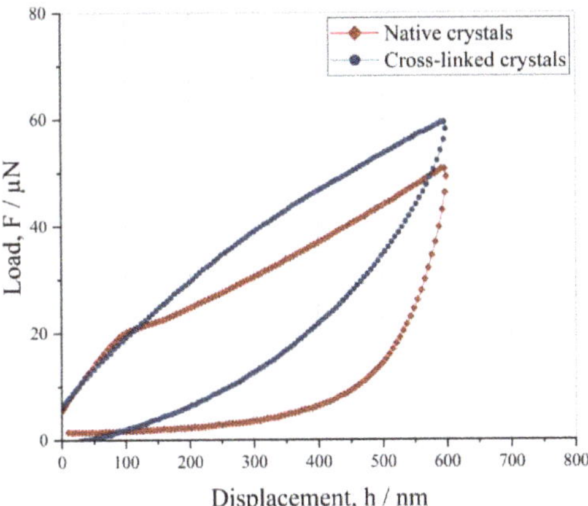

Figure 13. Force–displacement curve comparison of native and cross-linked HheG prismatic faces.

The hardness of the cross-linked HheG crystals in Figure 6 present higher hardness than the hardness of native crystals only at low penetration depths. This is caused by the strongly elastic recovery of the cross-linked crystals, especially at penetration depths of 200 and 300 nm. As a consequence, the slope of the unloading curve (P–h curve) is significantly smaller than that of native crystals. When calculating hardness according to the Oliver and Pharr method, the slope is used to determine the contact area—i.e., the smaller the slope is, the smaller the contact area will be. According to this definition, hardness refers to the maximum force obtained at the indentation in relation to the contact area. Therefore, the smaller the slope and contact area are, the greater the hardness will be.

Apart from an improved mechanical stability due to cross-linking, it is necessary for the CLECs to retain their catalytic activity for them to be used in biocatalytic reactions. To investigate the catalytic activity of the formed HheG CLECs, they were applied in opening the ring of cyclohexene oxide with azide as a nucleophile. The conversion data using different forms of HheG (soluble, CLECs) are presented in Figure 14.

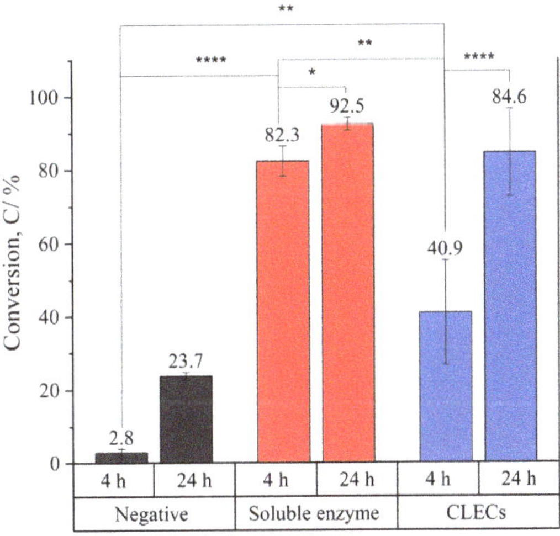

Figure 14. Epoxide ring opening of cyclohexene oxide with azide as a nucleophile after 4 and 24 h. A two-tailed t-test for two random samples was performed with a significance level of $p = 0.05$. Levels were determined as follows: <0.05 = *, <0.005 = **, <0.0005 = ***, and <0.00005 = ****.

Positive control reactions using 100 μg of soluble HheG yielded a mean conversion value of 90% after 24 h. This reaction was previously reported to reach product yields up to 98% within 24 h [4]. By comparison, the HheG CLECs achieved a conversion of 85% after 24 h (Figure 14). This result confirms the high residual activity of HheG CLECs compared to negative control reactions without the enzyme, which exhibited only 23% conversion after 24 h. In contrast, the HheG CLECs yielded only 41% conversion after 4 h of reaction, whereas a two times higher conversion rate (83%) was achieved using the soluble enzyme. This result indicates reduced HheG activity within the cross-linked enzyme crystal. This assumption is further underlined by the fact that the CLECs required a much longer reaction time to reach near completion of the reaction. Based on the literature, three main parameters seem to influence the activity of crystallized enzymes—crystal size, substrate size, and enzyme conformation [60]. The first two parameters are related to the diffusion limitations within the crystals and were addressed in several publications [22,63,64]. One possible solution to overcome diffusional limitations is the use of smaller enzyme crystals [65–67]. For example, Kasvinsky and Madsen reported that enzyme crystals of glycogen phosphorylase with a crystal size of 10 μm did not show diffusion limitations [65]. Mass transfer was also dependent on substrate concentration. Choosing a substrate concentration much higher than the respective K_m value of the enzyme seems to overcome diffusional limitations [10]. Besides mass transfer limitations, another explanation for reduced HheG activity could be the use of the cross-linker, glutaraldehyde. Glutaraldehyde reacts with the ε-amino groups of lysines but can also interact with the guanidyl groups of arginines [68] and tyrosines [68]. As Tyr165 and Arg169 are part of the catalytic triad of HheG [4], cross-linking of these residues with glutaraldehyde would also reduce enzyme activity and lead to reduced conversion when using CLECs compared to soluble enzymes. As our focus within this project was not to optimize CLEC activity but to investigate the influence of cross-linking on the mechanical behaviors of enzyme crystals, glutaraldehyde was used as a cross-linker. For the growth of crystals, a compromise had to be made between a crystal size that still showed sufficient catalytic activity and a size that could be reliably measured using a nanoindentation technique.

4. Conclusions

This project was undertaken to design robust and catalytically active CLECs and evaluate the differences in mechanical behavior between native and cross-linked enzyme crystals. Using the nanoindentation technique, the depth-dependent mechanical properties of anisotropic chemically treated and untreated enzyme crystals were investigated. This study provides several findings:

- The mechanical properties of native and cross-linked enzyme crystals, such as hardness and the fractions of elastic and plastic energy, decrease with an increase in penetration depth, meaning that they exhibit an indentation size effect (ISE).
- The mechanical properties are not dependent on the indentation rate, but the tendency toward adhesion increases at a slower displacement rate.
- Native crystals show anisotropic mechanical behaviors, where the basal face is more elastic than the prismatic one.
- The lower resistance of prismatic faces against plastic deformation seems to be caused by anisotropic crystallographic planes and intermolecular water.
- Cross-linking increases the fraction of the reversible indentation energy of prismatic and basal faces by approximately 68% and 8%, respectively.
- HheG CLECs cross-linked with glutaraldehyde showed high catalytic activity under a conversion time of 24 h. Due to diffusion limitations, a longer reaction time was required to complete the reaction compared to that when using a soluble enzyme.

Using a nanoindenter, we systematically identified the weakest spots on the HheG crystal surfaces and correlated those spots with the crystal structure. We also showed that the crystals' flaws due to cross-linking can be significantly reduced using glutaraldehyde. This is the first study reporting process-relevant properties, such as the elastic and plastic deformation energy of enzyme crystals, and extends our understanding of ways to enhance their mechanical performance under cross-linking while maintaining catalytic activity. This knowledge will be necessary if CLECs are to be successfully used in industrial processes.

Author Contributions: Conceptualization, investigation, and methodology, M.K.; data curation, M.K. and M.S.; funding acquisition, A.S. and C.S.; project administration, I.K. and C.S.; supervision, A.S., I.K., and C.S.; writing—original draft, M.K. and M.S.; writing—review and editing, M.K., A.S., I.K., and C.S. All authors have read and agreed to the published version of the manuscript.

Funding: This work was funded by the German Research Foundation (DFG), within the priority Programme DiSPBiotech (SPP 1934, SCHI 1265/3-1 and SCHA 1745/2-2).

Acknowledgments: We acknowledge financial support from the German Research Foundation and the Open Access Publication Funds of the Technische Universität Braunschweig.

Conflicts of Interest: The authors declare no conflict of interest.

References

1. Schallmey, A.; Schallmey, M. Recent advances on halohydrin dehalogenases-from enzyme identification to novel biocatalytic applications. *Appl. Microbiol. Biotechnol.* **2016**, *100*, 7827–7839. [CrossRef]
2. Hasnaoui-Dijoux, G.; Majerić Elenkov, M.; Lutje Spelberg, J.H.; Hauer, B.; Janssen, D.B. Catalytic promiscuity of halohydrin dehalogenase and its application in enantioselective epoxide ring opening. *ChemBioChem* **2008**, *9*, 1048–1051. [CrossRef] [PubMed]
3. Koopmeiners, J.; Halmschlag, B.; Schallmey, M.; Schallmey, A. Biochemical and biocatalytic characterization of 17 novel halohydrin dehalogenases. *Appl. Microbiol. Biotechnol.* **2016**, *100*, 7517–7527. [CrossRef] [PubMed]
4. Koopmeiners, J.; Diederich, C.; Solarczek, J.; Voß, H.; Mayer, J.; Blankenfeldt, W.; Schallmey, A. HheG, a Halohydrin Dehalogenase with Activity on Cyclic Epoxides. *ACS Catal.* **2017**, *7*, 6877–6886. [CrossRef]
5. Calderini, E.; Wessel, J.; Süss, P.; Schrepfer, P.; Wardenga, R.; Schallmey, A. Selective Ring-Opening of Di-Substituted Epoxides Catalysed by Halohydrin Dehalogenases. *ChemCatChem* **2019**, *11*, 2099–2106. [CrossRef]
6. An, M.; Liu, W.; Zhou, X.; Ma, R.; Wang, H.; Cui, B.; Han, W.; Wan, N.; Chen, Y. Highly α-position regioselective ring-opening of epoxides catalyzed by halohydrin dehalogenase from Ilumatobacter coccineus: A biocatalytic approach to 2-azido-2-aryl-1-ols. *RSC Adv.* **2019**, *9*, 16418–16422. [CrossRef]
7. Sternberg, J.A.; Geffken, D.; Adams, J.B.; Pstages, R.; Sternberg, C.G.; Campbell, C.L.; Moberg, W.K. Famoxadone: The discovery and optimisation of a new agricultural fungicide. *Pest. Manag. Sci.* **2001**, *57*, 143–152. [CrossRef]
8. Swaney, S.M.; Aoki, H.; Ganoza, M.C.; Shinabarger, D.L. The oxazolidinone linezolid inhibits initiation of protein synthesis in bacteria. *Antimicrob. Agents Chemother.* **1998**, *42*, 3251–3255. [CrossRef]
9. Heravi, M.M.; Zadsirjan, V.; Farajpour, B. Applications of oxazolidinones as chiral auxiliaries in the asymmetric alkylation reaction applied to total synthesis. *RSC Adv.* **2016**, *6*, 30498–30551. [CrossRef]
10. Tischer, W.; Kasche, V. Immobilized enzymes: Crystals or carriers? *Trends Biotechnol.* **1999**, *17*, 326–335. [CrossRef]
11. Guo, S.; Akhremitchev, B.B. Investigation of mechanical properties of insulin crystals by atomic force microscopy. *Langmuir* **2008**, *24*, 880–887. [CrossRef]
12. Tait, S.; White, E.T.; Litster, J.D. Mechanical Characterization of Protein Crystals. *Part. Part. Syst. Charact.* **2008**, *25*, 266–276. [CrossRef]
13. Koizumi, H.; Kawamoto, H.; Tachibana, M.; Kojima, K. Effect of intracrystalline water on micro-Vickers hardness in tetragonal hen egg-white lysozyme single crystals. *J. Phys. D Appl. Phys.* **2008**, *41*, 74019. [CrossRef]
14. Tachibana, M.; Kobayashi, Y.; Shimazu, T.; Ataka, M.; Kojima, K. Growth and mechanical properties of lysozyme crystals. *J. Cryst. Growth* **1999**, *198–199*, 661–664. [CrossRef]
15. Kishi, T.; Suzuki, R.; Shigemoto, C.; Murata, H.; Kojima, K.; Tachibana, M. Microindentation Hardness of Protein Crystals under Controlled Relative Humidity. *Crystals* **2017**, *7*, 339. [CrossRef]
16. Zamiri, A.; De, S. Modeling the mechanical response of tetragonal lysozyme crystals. *Langmuir* **2010**, *26*, 4251–4257. [CrossRef]
17. Koizumi, H.; Tachibana, M.; Kawamoto, H.; Kojima, K. Temperature dependence of microhardness of tetragonal hen-egg-white lysozyme single crystals. *Philos. Mag.* **2004**, *84*, 2961–2968. [CrossRef]
18. Rupp, B. *Biomolecular Crystallography: Principles, Practice, and Application to Structural Biology*; Garland Science: New York, NY, USA, 2010; ISBN 978-0815340812.
19. Yan, E.-K.; Lu, Q.-Q.; Zhang, C.-Y.; Liu, Y.-L.; He, J.; Chen, D.; Wang, B.; Zhou, R.-B.; Wu, P.; Yin, D.-C. Preparation of cross-linked hen-egg white lysozyme crystals free of cracks. *Sci. Rep.* **2016**, *6*, 34770. [CrossRef]
20. Zelinski, T.; Waldmann, H. Cross-Linked Enzyme Crystals (CLECs): Efficient and Stable Biocatalysts for Preparative Organic Chemistry. *Angew. Chem. Int. Ed. Engl.* **1997**, *36*, 722–724. [CrossRef]
21. St Clair, N.L.; Navia, M.A. Cross-linked enzyme crystals as robust biocatalysts. *J. Am. Chem. Soc.* **1992**, *114*, 7314–7316. [CrossRef]
22. Quiocho, F.A.; Richards, F.M. Intermolecular Cross Linking of a Protein in the Crystalline State: Carboxypeptidase-A. *Proc. Natl. Acad. Sci. USA* **1964**, *52*, 833–839. [CrossRef]

23. Govardhan, C.P. Crosslinking of enzymes for improved stability and performance. *Curr. Opin. Biotechnol.* **1999**, *10*, 331–335. [CrossRef]
24. Jegan Roy, J.; Emilia Abraham, T. Strategies in making cross-linked enzyme crystals. *Chem. Rev.* **2004**, *104*, 3705–3722. [CrossRef]
25. Yan, E.-K.; Cao, H.-L.; Zhang, C.-Y.; Lu, Q.-Q.; Ye, Y.-J.; He, J.; Huang, L.-J.; Yin, D.-C. Cross-linked protein crystals by glutaraldehyde and their applications. *RSC Adv.* **2015**, *5*, 26163–26174. [CrossRef]
26. Barbosa, O.; Ortiz, C.; Berenguer-Murcia, Á.; Torres, R.; Rodrigues, R.C.; Fernandez-Lafuente, R. Glutaraldehyde in bio-catalysts design: A useful crosslinker and a versatile tool in enzyme immobilization. *RSC Adv.* **2014**, *4*, 1583–1600. [CrossRef]
27. Morozov, V.N.; Morozova, T.Y. Viscoelastic properties of protein crystals: Triclinic crystals of hen egg white lysozyme in different conditions. *Biopolymers* **1981**, *20*, 451–467. [CrossRef]
28. Lee, T.S.; Turner, M.K.; Lye, G.J. Mechanical stability of immobilized biocatalysts (CLECs) in dilute agitated suspensions. *Biotechnol. Prog.* **2002**, *18*, 43–50. [CrossRef]
29. Kubiak, M.; Solarczek, J.; Kampen, I.; Schallmey, A.; Kwade, A.; Schilde, C. Micromechanics of Anisotropic Cross-Linked Enzyme Crystals. *Cryst. Growth Des.* **2018**, *18*, 5885–5895. [CrossRef]
30. Kubiak, M.; Mayer, J.; Kampen, I.; Schilde, C.; Biedendieck, R. Structure-Properties Correlation of Cross-Linked Penicillin G Acylase Crystals. *Crystals* **2021**, *11*, 451. [CrossRef]
31. Kubiak, M.; Storm, K.-F.; Kampen, I.; Schilde, C. Relationship between Cross-Linking Reaction Time and Anisotropic Mechanical Behavior of Enzyme Crystals. *Cryst. Growth Des.* **2019**, *19*, 4453–4464. [CrossRef]
32. Margolin, A.L. Novel crystalline catalysts. *Trends Biotechnol.* **1996**, *14*, 223–230. [CrossRef]
33. Oyen, M.L.; Cook, R.F. A practical guide for analysis of nanoindentation data. *J. Mech. Behav. Biomed. Mater.* **2009**, *2*, 396–407. [CrossRef] [PubMed]
34. Lusty, C.J. A gentle vapor-diffusion technique for cross-linking of protein crystals for cryocrystallography. *J. Appl. Crystallogr.* **1999**, *32*, 106–112. [CrossRef]
35. López-Jaramillo, F.J.; Moraleda, A.B.; González-Ramírez, L.A.; Carazo, A.; García-Ruiz, J.M. Soaking: The effect of osmotic shock on tetragonal lysozyme crystals. *Acta Crystallogr. Sect. D Struct. Biol.* **2002**, *58*, 209–214. [CrossRef] [PubMed]
36. Monsan, P.; Puzo, G.; Mazarguil, H. Étude du mécanisme d'établissement des liaisons glutaraldéhyde-protéines. *Biochimie* **1976**, *57*, 1281–1292. [CrossRef]
37. Solarczek, J.; Klünemann, T.; Brandt, F.; Schrepfer, P.; Wolter, M.; Jacob, C.R.; Blankenfeldt, W.; Schallmey, A. Position 123 of halohydrin dehalogenase HheG plays an important role in stability, activity, and enantioselectivity. *Sci. Rep.* **2019**, *9*, 5106. [CrossRef] [PubMed]
38. Oliver, W.C.; Pharr, G.M. Measurement of hardness and elastic modulus by instrumented indentation: Advances in understanding and refinements to methodology. *J. Mater. Res.* **2004**, *19*, 3–20. [CrossRef]
39. Schilde, C.; Burmeister, C.F.; Kwade, A. Measurement and simulation of micromechanical properties of nanostructured aggregates via nanoindentation and DEM-simulation. *Powder Technol.* **2014**, *259*, 1–13. [CrossRef]
40. Morozov, V.N.; Morozova, T.Y.; Myachin, E.G.; Kachalova, G.S. Metastable state of a protein crystal. *Acta Crystallogr. B Struct. Sci.* **1985**, *41*, 202–205. [CrossRef]
41. Nix, W.D.; Gao, H. Indentation size effects in crystalline materials: A law for strain gradient plasticity. *J. Mech. Phys. Solids* **1998**, *46*, 411–425. [CrossRef]
42. Suzuki, R.; Shigemoto, C.; Abe, M.; Kojima, K.; Tachibana, M. Analysis of slip systems in protein crystals with a triclinic form using a phenomenological macro-bond method. *CrystEngComm* **2021**, *23*, 3753–3760. [CrossRef]
43. Suzuki, R.; Kishi, T.; Tsukashima, S.; Tachibana, M.; Wako, K.; Kojima, K. Hardness and slip systems of orthorhombic hen egg-white lysozyme crystals. *Philos. Mag.* **2016**, *96*, 2930–2942. [CrossRef]
44. Aibara, S.; Suzuki, A.; Kidera, A.; Shibata, K.; Yamane, T.; DeLucas, L.J.; Hirose, M. The Crystal Structure of the Orthorhombic Form of Hen Egg White Lysozyme at 1.5 Angstroms Resolution. 2004. Available online: https://www.rcsb.org/structure/1VDQ (accessed on 22 June 2021).
45. Swadener, J.G.; George, E.P.; Pharr, G.M. The correlation of the indentation size effect measured with indenters of various shapes. *J. Mech. Phys. Solids* **2002**, *50*, 681–694. [CrossRef]
46. Kucharski, S.; Woźniacka, S. Size Effect in Single Crystal Copper Examined with Spherical Indenters. *Metall. Mater. Trans. A* **2019**, *50*, 2139–2154. [CrossRef]
47. Raut, D.; Kiran, M.S.R.N.; Mishra, M.K.; Asiri, A.M.; Ramamurty, U. On the loading rate sensitivity of plastic deformation in molecular crystals. *CrystEngComm* **2016**, *18*, 3551–3555. [CrossRef]
48. Buehler, M.J.; Ackbarow, T. Fracture mechanics of protein materials. *Mater. Today* **2007**, *10*, 46–58. [CrossRef]
49. Labonte, D.; Lenz, A.-K.; Oyen, M.L. On the relationship between indenation hardness and modulus, and the damage resistance of biological materials. *Acta Biomater.* **2017**, *57*, 373–383. [CrossRef]
50. Edwards, A.J.; Mackenzie, C.F.; Spackman, P.R.; Jayatilaka, D.; Spackman, M.A. Intermolecular interactions in molecular crystals: What's in a name? *Faraday Discuss.* **2017**, *203*, 93–112. [CrossRef]
51. Wang, C.; Sun, C.C. The landscape of mechanical properties of molecular crystals. *CrystEngComm* **2020**, *22*, 1149–1153. [CrossRef]
52. Reddy, C.M.; Rama Krishna, G.; Ghosh, S. Mechanical properties of molecular crystals—applications to crystal engineering. *CrystEngComm* **2010**, *12*, 2296. [CrossRef]

53. Dunitz, J.D. Weak Interactions in Molecular Crystals. In *Implications of Molecular and Materials Structure for New Technologies*; Howard, J.A.K., Allen, F.H., Shields, G.P., Eds.; Springer: Dordrecht, The Netherlands, 1999; pp. 175–184. ISBN 978-0-7923-5817-6.
54. Buehler, M.J. Atomistic modeling of elasticity, plasticity and fracture of protein crystals. *MRS Proc.* **2005**, *898*, 1003. [CrossRef]
55. Taylor, G.I. The mechanism of plastic deformation of crystals. Part I.—Theoretical: Geoffrey Ingram Taylor. *Proc. R. Soc. Lond. A* **1934**, *145*, 362–387. [CrossRef]
56. Taylor, G.I. The mechanism of plastic deformation of crystals. Part II.—Comparison with observations. *Proc. R. Soc. Lond. A* **1934**, *145*, 388–404. [CrossRef]
57. Suzuki, R.; Tachibana, M.; Koizumi, H.; Kojima, K. Direct observation of stress-induced dislocations in protein crystals by synchrotron X-ray topography. *Acta Mater.* **2018**, *156*, 479–485. [CrossRef]
58. Xia, W.; Dehm, G.; Brinckmann, S. Insight into indentation-induced plastic flow in austenitic stainless steel. *J. Mater. Sci.* **2020**, *55*, 9095–9108. [CrossRef]
59. Nanev, C.N. Brittleness of protein crystals. *Cryst. Res. Technol.* **2012**, *47*, 922–927. [CrossRef]
60. Margolin, A.L.; Navia, M.A. Protein Crystals as Novel Catalytic Materials. *Angew. Chem. Int. Ed. Engl.* **2001**, *40*, 2204–2222. [CrossRef]
61. DeSantis, G.; Jones, J.B. Chemical modification of enzymes for enhanced functionality. *Curr. Opin. Biotechnol.* **1999**, *10*, 324–330. [CrossRef]
62. Wong, S.S.; Wong, L.-J.C. Chemical crosslinking and the stabilization of proteins and enzymes. *Enzym. Microb. Technol.* **1992**, *14*, 866–874. [CrossRef]
63. Makinen, M.W.; Fink, A.L. Reactivity and cryoenzymology of enzymes in the crystalline state. *Annu. Rev. Biophys. Bioeng.* **1977**, *6*, 301–343. [CrossRef]
64. Westbrook, E.M.; Sigler, P.B. Enzymatic function in crystals of delta 5-3-ketosteroid isomerase. Catalytic activity and binding of competitive inhibitors. *J. Biol. Chem.* **1984**, *259*, 9090–9095. [CrossRef]
65. Kasvinsky, P.J.; Madsen, N.B. Activity of glycogen phosphorylase in the crystalline state. *J. Biol. Chem.* **1976**, *251*, 6852–6859. [CrossRef]
66. Spilburg, C.A.; Bethune, J.L.; Valee, B.L. Kinetic properties of crystalline enzymes. Carboxypeptidase A. *Biochemistry* **1977**, *16*, 1142–1150. [CrossRef]
67. Alter, G.M.; Leussing, D.L.; Neurath, H.; Vallee, B.L. Kinetic properties of carboxypeptidase B in solutions and crystals. *Biochemistry* **1977**, *16*, 3663–3668. [CrossRef]
68. Migneault, I.; Dartiguenave, C.; Bertrand, M.J.; Waldron, K.C. Glutaraldehyde: Behavior in aqueous solution, reaction with proteins, and application to enzyme crosslinking. *BioTechniques* **2004**, *37*, 790–802. [CrossRef]

Article

Crystal Structure of Nitrilase-Like Protein Nit2 from *Kluyveromyces lactis*

Chaewon Jin [1], Hyeonseok Jin [1], Byung-Cheon Jeong [2], Dong-Hyung Cho [3], Hang-Suk Chun [4], Woo-Keun Kim [4] and Jeong Ho Chang [1,5,*]

1 Department of Biology Education, Kyungpook National University, Daehak-ro 80, Buk-gu, Daegu 41566, Korea; jinchaewon178@gmail.com (C.J.); jinhs@postech.ac.kr (H.J.)
2 Department of Pharmacology, University of Texas Southwestern Medical Center, Dallas, TX 75390, USA; Byung-Cheon.Jeong@utsouthwestern.edu
3 School of Life Sciences, Kyungpook National University, Daegu 41566, Korea; dhcho@knu.ac.kr
4 Biosystem Research Group, Korea Institute of Toxicology, Daejeon 34114, Korea; hangsuk.chun@kitox.re.kr (H.-S.C.); wookkim@kitox.re.kr (W.-K.K.)
5 Department of Biomedical Convergence Science and Technology, Kyungpook National University, Daehak-ro 80, Buk-gu, Daegu 41566, Korea
* Correspondence: jhcbio@knu.ac.kr; Tel.: +82-53-950-5913; Fax: +82-53-950-6809

Abstract: The nitrilase superfamily, including 13 branches, plays various biological functions in signaling molecule synthesis, vitamin metabolism, small-molecule detoxification, and posttranslational modifications. Most of the mammals and yeasts have Nit1 and Nit2 proteins, which belong to the nitrilase-like (Nit) branch of the nitrilase superfamily. Recent studies have suggested that Nit1 is a metabolite repair enzyme, whereas Nit2 shows ω-amidase activity. In addition, Nit1 and Nit2 are suggested as putative tumor suppressors through different ways in mammals. Yeast Nit2 (yNit2) is a homolog of mouse Nit1 based on similarity in sequence. To understand its specific structural features, we determined the crystal structure of Nit2 from *Kluyveromyces lactis* (*Kl*Nit2) at 2.2 Å resolution and compared it with the structure of yeast-, worm-, and mouse-derived Nit2 proteins. Based on our structural analysis, we identified five distinguishable structural features from 28 structural homologs. This study might potentially provide insights into the structural relationships of a broad spectrum of nitrilases.

Keywords: Nit2; nitrilase superfamily; ω-amidase; *Kluyveromyces lactis*

Citation: Jin, C.; Jin, H.; Jeong, B.-C.; Cho, D.-H.; Chun, H.-S.; Kim, W.-K.; Chang, J.H. Crystal Structure of Nitrilase-Like Protein Nit2 from *Kluyveromyces lactis*. *Crystals* **2021**, *11*, 499. https://doi.org/10.3390/cryst11050499

Academic Editors: Abel Moreno, Kyeong Kyu Kim and T. Doohun Kim

Received: 13 March 2021
Accepted: 21 April 2021
Published: 1 May 2021

1. Introduction

The members of the nitrilase superfamily fulfill various biological roles, such as signaling molecule synthesis, vitamin and coenzyme metabolism, small-molecule detoxification, and posttranslational modifications [1]. Most family members share a common carbon–nitrogen (CN) hydrolase domain, found in plants, animals, fungi, as well as many prokaryotes [2,3]. The members contain a catalytic triad in their active sites consisting of highly conserved cysteine, glutamate, and lysine residues [2]. Despite the name of the superfamily, only one branch exhibits nitrilase activity, whereas others are known as acid amides, secondary amidases, ureas, and carbamates.

Based on a sequence analysis and domain fusion patterns, the members of the nitrilase superfamily could be classified into 13 branches [1], among which the nitrilase-like (Nit) protein is included in the 10th branch, originally categorized as Fhit-related tumor suppression proteins [4]. Two Nit subtypes, Nit1 and Nit2, could be found in the Nit-branch in mammals and yeast, which have ω-amidase activities [2,5]. The mammalian Nit1 protein is a homolog of the yeast Nit2 protein [2]. In mammals, Nit2 shows ω-amidase activity toward α-ketoglutaramate, whereas Nit1 has a weaker ω-amidase activity [6,7]. Nit2 acts to convert glutamine into α-ketoglutaramate, which replenishes α-ketoglutarate to the tricarboxylic acid (TCA) cycle. Some studies have shown that mammalian Nit1 has recently been assumed as an enzyme that functions in repairing metabolic damage. In the study, it was estimated that transaminase inadvertently produces dGSH and that Nit1 serves as a metabolite repair enzyme that converts dGSH to α-ketoglutaramate and cysteinylglycine [5]. The expression of the mammalian Nit1 protein (yeast Nit2 homolog) is separated from that of Fhit. However, the NitFhit Rosetta stone protein could be found in invertebrates [8]. This suggests that Nit1 and Fhit are likely to be involved in the same signaling or metabolic pathway, and their activities could be additive.

Several reports suggest that Nit1 and Nit2 are potent tumor suppressors [4,9–13]. Nit1-knockout mice have accelerated cell proliferation with increased expression of cyclin D1, enhanced survival of cells having DNA damage, and increased incidence of *N*-nitrosomethylbenzylamine-induced tumors [12,14]. However, Nit1 overexpression induces both decreased cell viability and increased caspase-3-dependent apoptosis [14]. Thus, the overexpression of Nit2 in HeLa cells leads to a decrease in cell proliferation and growth through induction of G2 arrest rather than by apoptosis. This also indicates that Nit2 can potentially play a role in tumor suppression [11]. Although Nit1 is involved in tumor suppression in multiple ways, it remains unclear whether these roles are related to its catalytic function [5]. Moreover, a mutation on the conserved active-site cysteine to alanine did not alter the ability of Nit1 to promote apoptosis in human cells.

In this study, to understand the specific structural features of Nit2, we determined the crystal structure of Nit2 from *Kluyveromyces lactis* (*Kl*Nit2) at 2.2 Å resolution and compared the three-dimensional structures of nitrilase superfamily members. Based on the structural analysis of Nit2, we managed to distinguish five structural features including the C-terminal region. In addition, we performed the sequence-based phylogenetic analysis of *Kl*Nit2 homologous structures. This study might provide insights into the structural and evolutionary relationships of a broad spectrum of nitrilases.

2. Materials and Methods

2.1. Cloning of Nit2 from Kluyveromyces lactis

The Nit2 gene was amplified from a *Kluyveromyces lactis* genomic DNA library through a polymerase chain reaction (PCR), as described previously [15]. Briefly, the amplified fragment was digested with the restriction enzymes NheI and XhoI (R016S and R0075, respectively, Enzynomics, Daejeon, Korea), and ligated into pET28a vectors using T4 DNA ligase (M0202S, Roche, Basel, Germany). The plasmid was then transformed into *Escherichia coli* (*E. coli*) strain DH5α, and the transformants were confirmed using a colony PCR. All oligonucleotides used in the study were purchased from Cosmogenetech (Seoul, Korea).

2.2. Purification of the Recombinant KlNit2 Protein

The plasmids encoding *Kluyveromyces lactis* Nit2 (*Kl*Nit2) proteins were transformed into the BL21 (DE3) Star *E. coli* strain. Cells were grown and induced with 0.3 mM isopropyl β-D-1-thiogalactopyranoside (IPTG; 420322, Calbiochem, Sigma-Aldrich, St. Louis, MO, USA) for 16 h at 20 °C in LB medium (L4488, MBcell, Seoul, Korea). The harvested cells were disrupted by ultrasonication. The lysate was bound to Ni-NTA agarose (30230, Qiagen, Hilden, Germany) for 90 min at 4 °C. After washing with buffer A (200 mM NaCl, 50 mM Tris, pH 8.0) containing 20 mM imidazole (I5513, Sigma-Aldrich, St. Louis, MO, USA), the bound proteins were eluted with 250 mM imidazole in buffer A. Size exclusion chromatography (SEC) was performed using HiPrep 16/60 Sephacryl S-300 HR (17116701, GE Healthcare, Chicago, IL, USA). The buffer used for SEC contained 150 mM NaCl, 2 mM dithiothreitol (DTT; 233155, Calbiochem, Sigma-Aldrich, St. Louis, MO, USA), and 20 mM Tris, pH 7.5. Following SEC, the protein was stored at −80 °C pending crystallization trials.

2.3. Crystallization

All crystallization trials were performed at 4 °C using either sitting-drop or hanging-drop vapor diffusion methods. The crystals for *Kl*Nit2 with the N-terminal His_6 tag grew within a day in drops containing equal volumes (1 μL) of protein and reservoir solutions (26% *w/v* polyethylene glycol (PEG 3350), 0.3 M ammonium tartrate dibasic). To improve the crystals, additional screening was performed using an additive (HR2-428, Hampton Research, Aliso Viejo, CA, USA) and a detergent (HR2-406, from Hampton Research, Research, Aliso Viejo, CA, USA) screening kit. Prior to flash cooling all crystals in liquid nitrogen, 30% glycerol was added to the reservoir solutions as a cryoprotectant.

2.4. Data Collection and Structure Determination

Diffraction datasets were collected at 100 K on beamline 5C of the Pohang Accelerator Laboratory (PAL, Pohang, Korea) using a Quantum 315 CCD detector (San Jose, CA, USA). Data were processed using the HKL-2000 software suite. The crystals of *Kl*Nit2 with the *N*-terminal His_6 tag belonged to the space group *C2* and diffracted to a resolution of 2.2 Å. The crystal structures were solved by molecular replacement methods using the *Phaser-MR* in the PHENIX crystallographic software, version 1.9 (PHENIX, Lawrence Berkeley Laboratory, Berkeley, CA, USA) [16]. The Nit2 from *Saccharomyces cerevisiae* (PDB ID: 4HG3) was used for the initial search model. The model building was performed using the Wincoot program [17]. The structural models were refined using the Phenix.refine program in the package of PHENIX program. All structural figures were visualized by the graphics program PyMOL (Schrödinger Inc., New York, NY, USA). The details of data collection and applied statistics are provided in Table 1.

Table 1. Data collection and refinement statics for *Kl*Nit2.

	*Kl*Nit2
Data collection	
Space group	C2
Cell dimensions	
a, b, c (Å)	79.0, 211.1, 89.4
α, β, γ (°)	90, 112.1, 90
Resolution range (Å)	50.0–2.2 (2.28–2.20) [1]
R_{merge} (%) [2]	8.9 (52.1)
$I/\sigma I$	17.0 (2.6)
Total reflections	392,121
Unique reflections	68,506 (6799)
Completeness (%)	99.9 (99.9)
Redundancy	5.7 (5.6)
$CC_{1/2}$	99.1 (86.2)
Structure refinement	
Resolution range (Å)	45.7–2.2
No. of reflections	65,198
R_{work} [3] / R_{free} (%) [4]	16.8/22.3
No. atoms	
Protein	9700
Water	543
RMSD	
Bond length (Å)	0.009
Bond angle (°)	0.941
Planar group (Å)	0.005
Chiral volume (Å^3)	0.057
Average *B*-factor (Å^2)	39.4
Protein	39.2
Solvent	43.3
Ramachandran plot (%)	
Favored region	97.2
Allowed	2.4
Disallowed	0.4
PDB code	7ELF

[1] The numbers in parentheses are statistics from the highest-resolution shell. [2] $R_{merge} = \Sigma\, |I_{obs} - I_{avg}| / I_{obs}$, where I_{obs} is the observed intensity of individual reflection, and I_{avg} is averaged over symmetry equivalents. [3] $R_{work} = \Sigma\, ||F_o| - |F_c|| / \Sigma\, |F_o|$, where $|F_o|$ and $|F_c|$ are the observed and calculated structure factor amplitudes, respectively. [4] R_{free} was calculated using 5% of the data.

3. Results

3.1. Overall Structure of KlNit2

The *Kl*Nit2 protein showed high homology with its homologous proteins such as *Saccharomyces cerevisiae* Nit2 (ScNit2; sequence identity, 60.32%), *Caenorhabditis elegans* Nit-Fhit (*ce*NitFhit; sequence identity, 33.44%), and *Mus musculus* Nit2 (*Mm*Nit2; sequence identity, 29.84%) (Figure 1A).

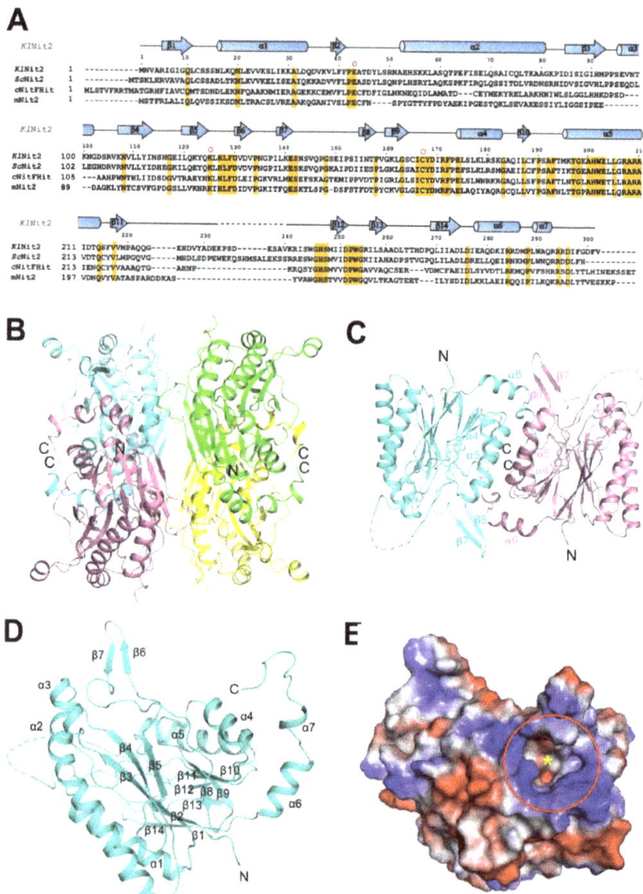

Figure 1. Overall structure of KlNit2. (**A**) Sequence alignment of KlNit2, ScNit2, mNit2, and the Nit region of cNitFhit. The secondary structure elements of KlNit2 are highlighted at the top of the sequences. Black dots above the sequences indicate every tenth residue in KlNit2. Identical residues within KlNit2 and its homologs are highlighted in yellow and the key residues of the active site are marked by a red circle on the sequences. (**B**) The tetrameric structure of KlNit2 is shown in a ribbon diagram. (**C**) The dimer structure of KlNit2. One monomer is shown in cyan; the other is colored light pink. The dashed lines show the regions that were not visible in the electron density map. (**D**) The monomer structure of KlNit2. The dashed line shows the region that was not visible in the electron density map. (**E**) The electrostatic surface model of the KlNit2 monomer. Red and blue represent the negatively and positively charged surfaces, respectively. The active site is highlighted in a red circle. The figures were generated using the graphics program PyMOL (Schrödinger Inc., New York, NY, USA).

While the *Kl*Nit2 protein forms a dimer in solution [15], four molecules exist in the asymmetric unit (Figure 1B). Dimerizations between subunits I and III as well as subunits II and IV are mediated by their α4 and α5 helices. Although both dimers can be stabilized by β9–α4 and β12–β13 loops, such interactions do not seem to be strong for each dimer. More stable dimers are generated by subunit association between neighboring symmetry-related molecules (Figure 1C). The dimeric interface is composed of α4, α5, α6, β6, and β7 that contribute to extensive interactions, as shown in a previous study of *Saccharomyces cerevisiae* Nit2 (*Sc*Nit2) [18]. The overall structure of *Kl*Nit2 has been described as an α–β–β–α architecture, similar to the structure of other nitrilase superfamily members [3,18–20]. Each subunit is composed of 7 α-helices (α1–α7) and 14 β-strands (β1–β14), 2 central β-sheets consisting of 12 β-strands (except β6 and β7), surrounded by 5 α-helices (except for α6 and α7) (Figure 1D). In addition, protruded β6 and β7 strands (β6–β7 hairpin) are positioned to cover the vicinity of a deep pocket containing a Lys–Cys–Glu triad, which plays an important role in the amidase activity of the protein (Figure 1D,E).

3.2. Active Site

The active site of *Kl*Nit2 is located in the deep pocket with a structure that closely resembles that of the other nitrilase superfamily members, such as mammalian Nit2 [3], yeast Nit2 [18], worm NitFhit Nit domain [8], bacterial Nit2 [20], and archaeal Nit2 [19] (Figure 2A). We observed that the β6 and β7 strands are localized in the vicinity of the active site. Thus, in the dimeric structure, the α6 and α7 helices from the neighboring subunit could be found near the active site. The distance between the catalytic sulfur of Cys167 and the oxygen atom of Glu43 is within the range of a hydrogen bond (Figure 2B). However, the distance between the sulfur atom of Cys167 and the N^{ε} atom of Lys125 is 6.09 Å that is slightly bigger than that observed in the yeast Nit2 structure [18]. It might be due to the ligand α-ketoglutarate binding the yeast Nit2 (Figure 3A). Thus, a thioester bond could be formed between the γ-carboxyl of the substrate and the –SH part of the active-site cysteine in the yeast Nit2. The distance between the oxygen of Glu43 and the N^{ε} atom of Lys125 is 4.63 Å, which is similar to that observed in the yeast Nit2 structure. The catalytic residues are well-conserved in the active site (Figure 1A). It is known that the catalytic triad with Ala192 and Tyr167 helps to bind a ligand intermediate [18] (Figure 2C). Thus, the residues Arg171, Thr194, and Thr197 immobilize the α-carboxyl group ligand. The aromatic residues Phe129 and Phe193 maintain a hydrophobic environment.

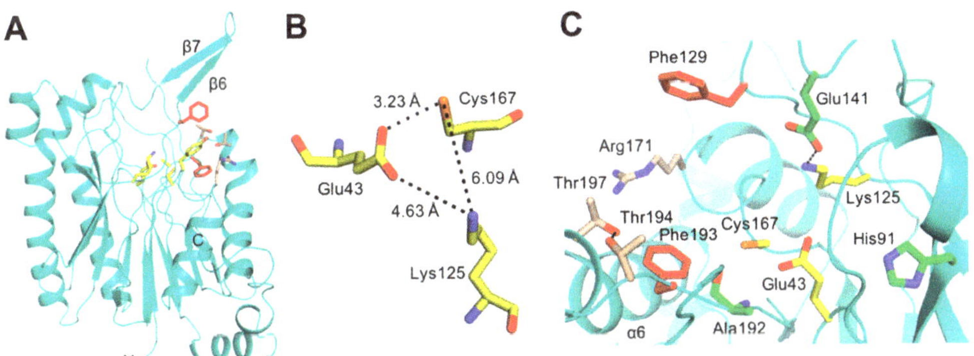

Figure 2. KlNit2 active site. (**A**) Monomeric structure including the active site. (**B**) The catalytic triad of the active site with distances shown between the key catalytic atoms. (**C**) The key residues of the active site. The catalytic triad (Glu43–Lys125–Cys167) is shown in yellow. The residues (Arg171, Thr194, and Thr197) that immobilize the α-carboxyl are shown in the color wheat. The residues (Phe129 and Phe193) that maintain a hydrophobic environment are shown in red. Figures were generated by the graphics program PyMOL (Schrödinger Inc., New York, NY, USA).

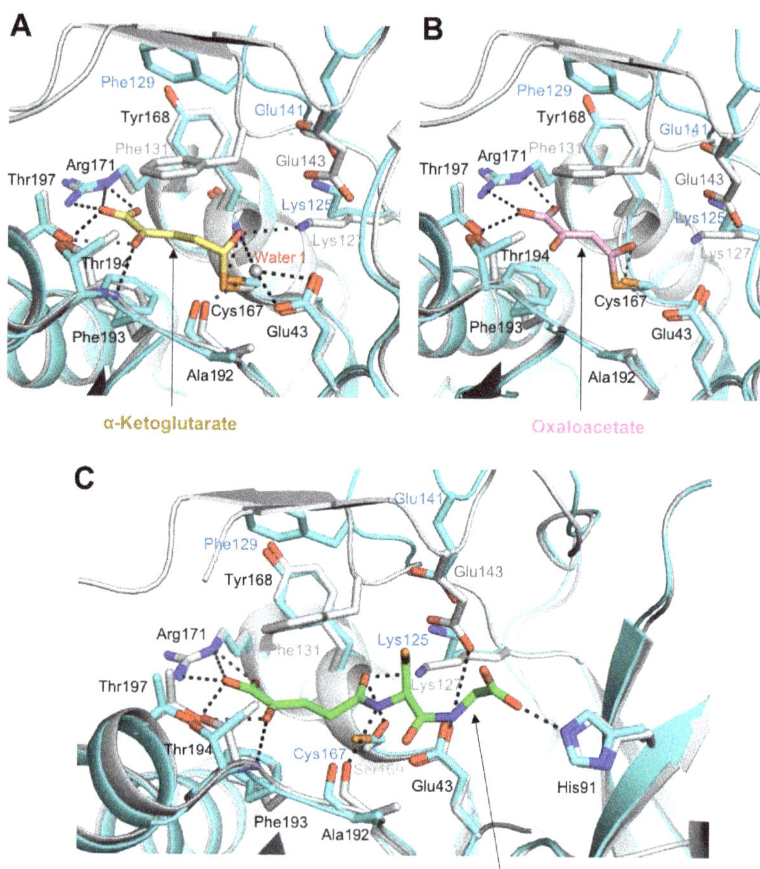

Figure 3. Superposition of the KlNit2 and ScNit2 ligand complex structures. (**A**) Detailed view of the interaction between α-ketoglutarate and overlaid KlNit2 with Nit2 from *Saccharomyces cerevisiae* (ScNit2; PDB code, 4HG3). Residues from KlNit2 with ScNit2 are shown in cyan and gray, respectively; α-ketoglutarate is shown in yellow. Water molecule 1 is shown as a gray sphere. (**B**) Detailed view of the interaction between oxaloacetate (OAA) and overlaid KlNit2 with ScNit2 (PDB code, 4HG5). The residues from KlNit2 with ScNit2 are shown as in Figure 3A. The OAA is shown in pink. (**C**) Detailed view of the interaction between a peptide-like ligand and overlaid KlNit2 with ScNit2 (PDB code, 4HGD). The residues from KlNit2 with ScNit2 are shown as in Figure 3A. The ligand is shown in green. The figures were generated using the graphics program PyMOL (Schrödinger Inc., New York, NY, USA).

3.3. Ligand-Binding Modes

To understand the ligand-binding mechanism of *Kl*Nit2, we compared the structures of *Sc*Nit2 in complexes with two products: α–ketoglutarate (α-KG) (PDB code: 4HG3) and oxaloacetate (OAA) (PDB code: 4HG5) [18]. A structure of the *Sc*Nit2 C169S mutant (PDB code: 4HGD) was also reported in complex with an unknown endogenous glutathione-like ligand, *N*-(4-carboxy-4-oxo-butanoyl)-l-cysteinylglycine (KGT) in the active site. We superimposed these previously published structures on that of *Kl*Nit2. While most residues in the active site are well aligned with those of the corresponding residues in *Sc*Nit2, several residues showed different conformations. The positions of Phe129 and Glu141 in *Kl*Nit2, corresponding to the Phe131 and Glu143 of *Sc*Nit2, significantly changed toward the ligands (Figure 3A–C). The Phe131 with *Sc*Nit2 Phe195 covers the thioester region of the ligands in sandwich positions. However, Glu143 does not interact with the two products α-KG and OAA, whereas it interacts with the amide group of KGT (Figure 3C), which might be due to the large movement of the β6–β7 hairpin driven by Phe129 (Phe131 of *Sc*Nit2) upon ligand binding (see Section 3.4). In the three ligand-binding structures of *Sc*Nit2, Arg173, Thr196, and Thr199 mainly contribute to stabilize the α-keto acidic group of the ligands by hydrogen bonds (Figure 3A–C). However, a water molecule mediates interactions between the active-site residues (Glu45, Lys127, and Ala194) and the carbonyl group of α-KG (Figure 3A). Interestingly, no notable interaction could be found in the OAA complex structure (Figure 3B). In the KGT complex structure, His93 and Glu143 established additional interactions with the amide and carboxylic acid groups (Figure 3C). Regarding the active site environment, the ligand-binding modes of *Kl*Nit2 might be similar to those of *Sc*Nit2.

3.4. Structural Comparison between KlNit2 and Its Homologs

To identify structurally varied regions of *Kl*Nit2, we selected three available homologous structures and superimposed: *Sc*Nit2 (PDB code, 4H5U), *Mm*Nit2 (PDB code, 2W1V), and *Ce*NitFhit (PDB code, 1EMS) [3,8,18]. Based on the DALi server output, *Sc*Nit2 structure exhibited the most similarity to that of *Kl*Nit2 with a *Z*-score of 45.9 and a root mean square deviation (r.s.m.d.) value of 1.3 Å [21] (Table 2).

Table 2. Structural similarity comparison of homologous structures using Dali [1].

Proteins	Species	Z-Score	RMSD (Å)	Identity (%)	Cα	PDB Code
Nit2	*S. cerevisiae*	46.1	1.3	63	292	4HG3
NitFhit	*C. elegans*	39.1	1.5	38	412	1EMS
Nit2	*Mus musculus*	33.9	2.2	32	274	2W1V
Nit3	*S. cerevisiae*	32.3	2.1	30	271	1F89
Putative carbon–nitrogen family hydrolase	*Staphylococcus aureus*	32.1	2.1	22	268	3P8K
Hypothetical protein Ph0642	*Pyrococcus horikoshii*	32.0	2.1	23	262	1J31
Hyperthermophilic nitrilase	*Pyrococcus abyssi*	31.5	2	22	261	3IVZ
NitN Amidase	*Nesterenkonia* sp. AN1	31.5	2	21	155	5JQN
Amidase	*Nesterenkonia* sp. 10004	31.4	1.9	21	255	5NYB
Carbon–nitrogen hydrolase	*Helicobacter pylori*	31.1	1.9	24	293	6MG6
N-carbamoyl-D-amino acid amidohydrolase	*Agrobacterium* sp.	31	2.5	30	303	1ERZ
Amidase	*Bacillus* sp.	31	3	17	339	4KZF
N-carbamoyl-D-amino-acid amidohydrolase	*Rhizobium radiobacter*	30.9	2.5	21	302	2GGK
Medicago truncatula N-carbamoylputrescine amidohydrolase (MtCPA)	*Medicago truncatula*	30.9	2.9	22	292	5H8I
Aliphatic amidase	*Pseudomonas aeruginosa*	30.7	2.4	18	341	2UXY
Bacillus cereus formamidase (BceAmiF)	*Bacillus cereus*	30.7	2.5	17	277	5H3O
Putative Nit protein	*Xanthomonas campestris pv.*	29.9	2.5	20	265	2E11
Glutamine-dependent NAD$^+$ synthetase	*Mycobacterium tuberculosis*	29.9	2.2	19	650	3SZG
Pyrimidine-degrading enzyme	*Drosophila melanogaster*	29.4	2.4	21	379	2VHI
Formamidase AmiF	*Helicobacter pylori*	29.1	2.2	19	317	2E2L
Glutamine-dependent NAD$^+$ synthetase	*Burkholderia thailandensis*	28.7	2.3	18	540	4F4H
β-Ureidopropionase	*Homo sapiens*	28.4	2.6	19	332	6FTQ
Nit6803	*Synechocystis* sp.	28.3	2.5	20	287	3WUY
Glutamine-dependent NAD$^+$ synthetase	*Acinetobacter baumannii*	28.2	2.4	19	526	5KHA
Nit4	*Arabidopsis thaliana*	28.1	2.1	23	289	6I00
Nh3-dependent NAD$^+$ synthetase	*Streptomyces avermitilis*	27.6	2.3	19	568	3N05
Glutamine-dependent NAD$^+$ synthetase	*Cytophaga hutchinsonii*	27.4	2.1	20	580	3ILV
Vanin-1	*Homo sapiens*	26.6	3.5	16	462	4CYF

[1] This server computes optimal and suboptimal structural alignments between two protein structures using the DaliLite-pairwise option. http://ekhidna.biocenter.helsinki.fi/dali/ (accessed on 21 April 2021).

Next, the structures of *Ce*NitFhit and *Mm*Nit2 exhibited similarity with a *Z*-score of 39.1 and an r.s.m.d. value of 1.5 Å and *Mm*Nit1with a *Z*-score of 33.9 and an r.s.m.d. value of 2.2 Å, respectively. Overall, we identified several regions, while the four structures shared two central β-sheets surrounded by five α-helices (Figure 4).

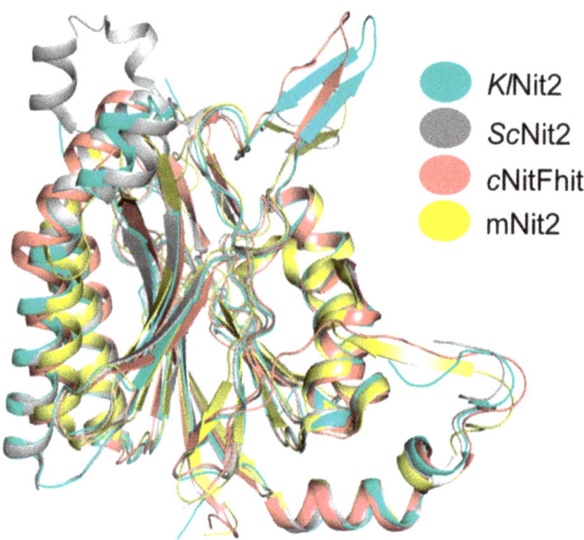

Figure 4. Comparison between *Kl*Nit2 and its homologs. Structural superimposition of *Kl*Nit2 (cyan) with *Sc*Nit2 (gray), mNit2 (yellow) and with the Nit region of *c*NitFhit (salmon). Figures were generated using the graphics program PyMOL (Schrödinger Inc., New York, NY, USA).

When comparing *Kl*Nit2 with *Sc*Nit2, the β6/7 hairpin structure, which covers the substrate tunnel, shows different angles. The superimposition of the monomer structures of each protein showed that β6/7 hairpin differs by 48.99° (Figure 5A, left panel). Therefore, the β6/hairpin of *Kl*Nit2 moved 16.70 Å outward of the *Sc*Nit2 β6/hairpin. Another difference is that there are two more β strands of *Sc*Nit2. In the case of *Sc*Nit2, two more β strands can be found between β11 and β12 of *Kl*Nit2 (Figure 5A, right panel). Considering the *Sc*Nit2 structure, we found two major distinct regions due to a crystallographic packing effect (Figure 6).

We observed no significant change of the β6–β7 hairpin in *Ce*NitFhit (Figure 5B, left panel). However, we identified no α3 helix, but short-turn loop due to sequence variation. In addition, the loop between β11 and β12 was significantly shorter than that of *Kl*Nit2 (Figure 5B, right panel). While the length of the loop was almost 30 residues in the KlNit2, only 15 residues could be identified in *Ce*NitFhit. Based on the sequence alignment, this common region varied the most among the three KlNit2 homologs (Figure 1A).

Several locally distinct regions were found in the *Mm*Nit2 structure. First, similar to the *Sc*Nit2 structure, the β6–β7 hairpin rotated toward the active site approximately 47° compared to the *Kl*Nit2 structure (Figure 5C, left panel). Second, similar to the structure of *Ce*NitFhit, the length of the loop between β11 and β12 was shorter than that in *Kl*Nit2 (Figure 5C, right panel). Among the four homologs, *Mm*Nit2 exhibited the shortest loop (Figure 1A). Third, an additional β-strand was found in its C-terminal end (Figure 5C, right panel). Fourth, the long *Kl*Nit2 α2 helix was divided into two helices in the *Mm*Nit2 homolog (Figure 5C, left panel). Finally, similar to the *Ce*NitFhit structure, we could observe no α3 helix but a short-turn loop containing four residues (Figure 5C, left panel).

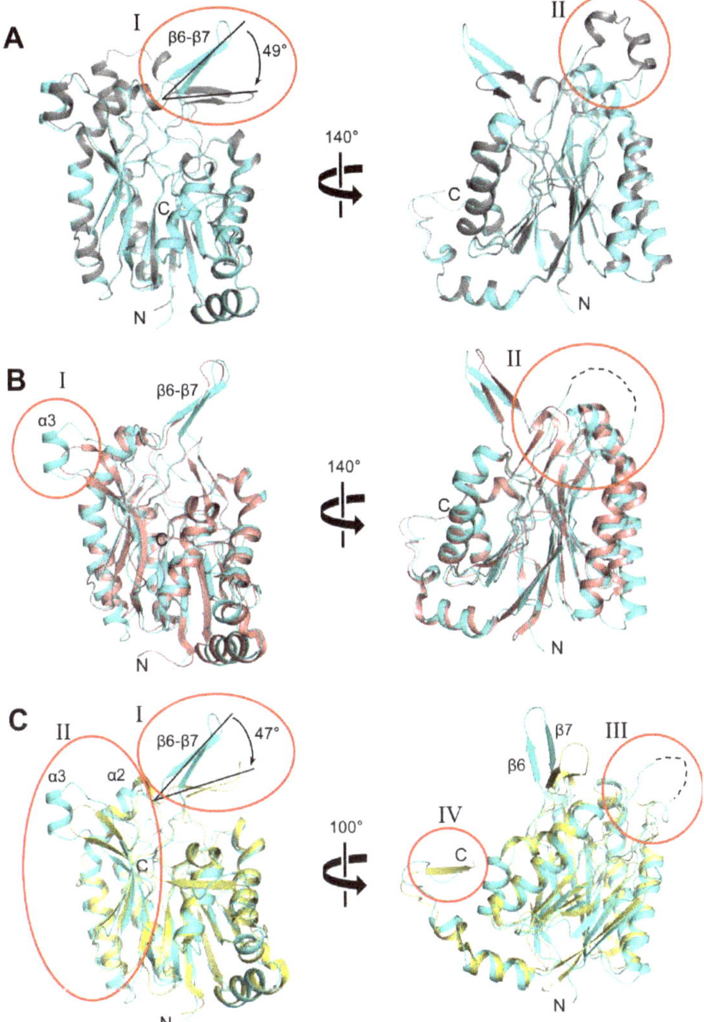

Figure 5. Structural comparison of Nit2 homologs. (**A**) The overall structures of KlNit2 and ScNit2 (PDB code, 4H5U) are depicted in the left panel in cyan and gray, respectively. The right panel shows the view of a 140° rotation along the *Y*-axis from the orientation compared to the left panel. The two different structural regions (I, II) are indicated by red circles with annotations. (**B**) The left panel shows the overall structures of KlNit2 and NitFhit from *Caenorhabditis elegans* (cNitFhit; PDB code, 1EMS) in cyan and salmon, respectively. The right panel highlights the view of a 140° rotation along the *Y*-axis compared to the orientation shown in the left panel. The two different structural regions (I, II) are indicated by red circles with annotations. (**C**) The overall structures of KlNit2 and mouse Nit2 (MmNit2, PDB code, 2W1V) are depicted in the left panel in cyan and yellow, respectively. The right panel shows the view of a 100° rotation along the *Y*-axis compared to the orientation shown in the left panel. The four different structural regions (I–IV) are indicated by red circles with annotations. The figures were generated using the graphics program PyMOL (Schrödinger Inc., New York, NY, USA).

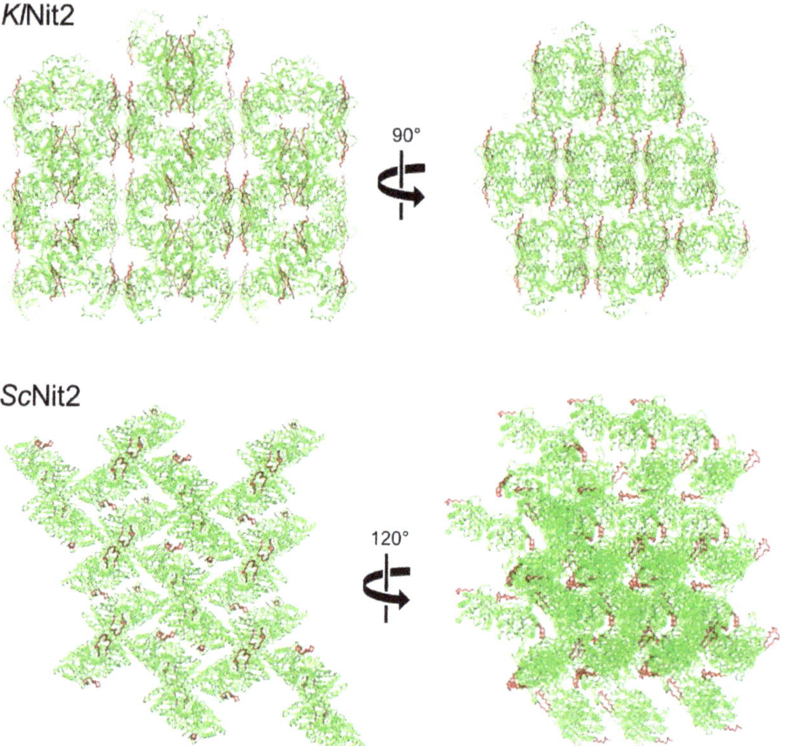

Figure 6. Crystallographic packing of Nit2 proteins. Comparison of crystallographic packing of KlNit2 and ScNit2. The figures were generated using the graphics program PyMOL (Schrödinger Inc., New York, NY, USA).

3.5. Comparison of the Active Site

The sequence alignment of the *Kl*Nit2 homologs showed that most residues exhibiting a relevant activity were highly conserved (Figure 1A). The superimposed structures of *Kl*Nit2 with those of *Sc*Nit2, *Ce*NitFhit, and *Mm*Nit2 showed a good overlay in the case of most residues (Figure 7A–C). The overlaid structures of *Kl*Nit2 and *Sc*Nit2 indicated that most residues interacting with the ligands matched well. However, the positions of Phe131, Glu143, and Lys127 in *Sc*Nit2 were located near the catalytic cysteine residue (Figure 7A). As mentioned previously, this effect might be due to conformational difference in the β6–β7 hairpin. Likewise, when superimposing the structure of *Kl*Nit2 and *Ce*NitFhit, most residues matched well, and even Phe129, Glu141, Lys125 were mostly similar as well (Figure 7B). The overlaid structures of *Kl*Nit2 and *Mm*Nit2 showed several different residue conformations (Figure 7C). Like the previous sequences, most residues were the same, but His91 in *Kl*Nit2 was altered to Pro124 in *Mm*Nit2. Although it is not involved in product binding, it might contribute to substrate specificity based on the structure of the ScNit2–KGT complex (Figure 3C). In addition, the conformation of the β2–α2 loop, including Ser49 of *Kl*Nit2, changed in *Mm*Nit2, containing Tyr87 oriented toward the catalytic cysteine. Phe129 and Glu141 of *Kl*Nit2 were also different in *Mm*Nit2 similar to that in *Sc*Nit2. As a result, these differences might contribute to the smaller active site of *Mm*Nit2 than that of *Kl*Nit2.

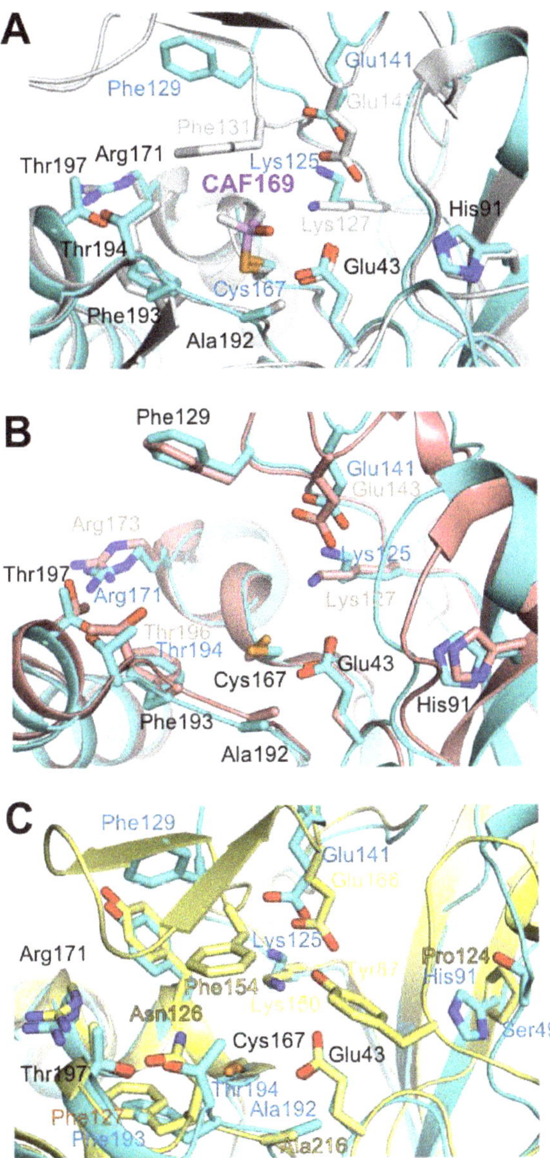

Figure 7. Comparison of the active sites. (**A**) Overlaid active sites of KlNit2 (cyan) and ScNit2 (gray). Residues with less varied conformations are numbered based on KlNit2 in black. Residues in distinct conformations are labeled by correspondence structures and colors. The S-dimethylarsinoyl-cystein from ScNit2 is in purple with an annotation. (**B**) Overlaid active sites of KlNit2 (cyan) and cNitFhit (salmon). Residues with less varied conformations are numbered based on KlNit2 in black. Residues in distinct conformations are labeled by correspondence structures and colors. (**C**) Overlaid active sites of KlNit2 (cyan) and MmNit2 (yellow). Residues with less varied conformations are numbered based on KlNit2 in black. Residues in distinct conformations are labeled by correspondence structures and colors. The figures were generated using the graphics program PyMOL (Schrödinger Inc., New York, NY, USA).

3.6. The Structural Features of the KlNit2-Homologous Proteins

In order to expand our knowledge concerning the structural features in the nitrilase protein family, we searched for *Kl*Nit2 structural homologs using the DALI server [21]. Based on the outputs, we could select 28 proteins under a *Z*-score of 26.6 and an r.s.m.d. value of 3.5 Å (Table 2). The proteins could be divided into ten categories according to their functions. The 22 proteins exhibited 9 functions: nitrilase, NAD$^+$ synthetase, nitrilase-like protein (Nit), aliphatic amidase, formamidase, amidohydrolase, beta-ureidopropionase, pyrimidine degrading enzyme, and pantetheinase. Moreover, the 6 proteins also exhibited uncharacterized functions. Following the comparison of the overall structures of 28 proteins, we found 5 specific local conformations. Based on the *Kl*Nit2 structure, the five distinct structural features include an α2 helix, a β3–α3–β4 motif, a β6–β7 hairpin, β11–β12 strands not visible in *Kl*Nit2, and a C-terminal part (Figure 8).

The structural feature of the α2 helix was divided into three categories (Figure 8A). While *Kl*Nit2 contains a single α2 helix, partially bent at approximately 132°, the other two types have fragmented helices. The single bent helix feature is widely present among the Nit proteins. The second type showed a deep bent helix of approximately 74° at the point of two turns from its N-terminal part in *Nsp*AN1NitN_Amidase. This feature could also be found in most NAD$^+$ synthetases and uncharacterized proteins from *Nesterenkonia* sp. 10004. The other type exhibited two fully separated helices connected by a loop as shown in *Ph*Hp_PH0642.

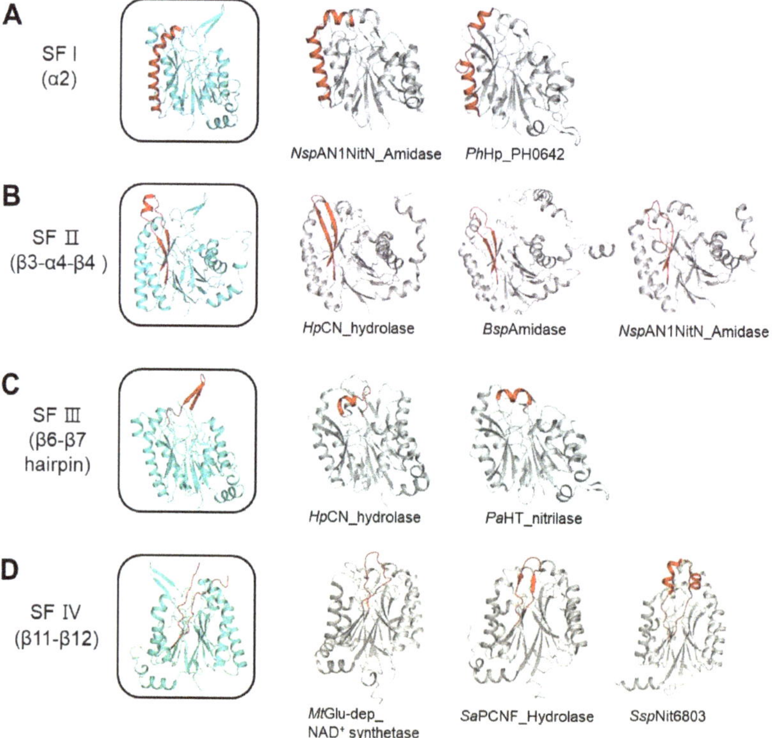

Figure 8. *Cont.*

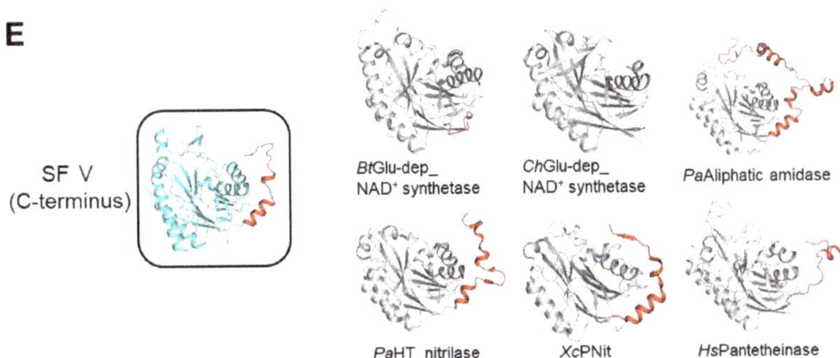

Figure 8. Five distinct structural features. (**A**) The representative structures containing structural feature I (SF I) according to the α2 helix. The KlNit2 structure is colored in cyan within a black rectangular box. The two structural homologs with a different conformation of the α2 helix are shown in gray. The α2 helix is shown in red. The two homologous structures indicate the following: NspAN1NitN_Amidase: NitN amidase from *Nesterenkonia* sp. AN1 (PDB code, 5JQN); PhHp_PH0642: Hypothetical Protein PH0642 from *Pyrococcus horikoshii* (PDB code, 1J31). (**B**) The representative structures containing structural feature II (SF II) according to the β3–α4–β4 motif based on KlNit2. The KlNit2 structure is colored in cyan within a black rectangular box. The three structural homologs with conformations different from β3–α4–β4 are shown in gray. The β3–α4–β4 are shown in red. The three homologous structures indicate the following: HpCN hydrolase: carbon–nitrogen (CN) hydrolase from *Helicobacter pylori* (PDB code, 6MG6), BspAmidase: Amidase from *Bacillus* sp. (PDB code, 4KZF), NspAN1NitN_Amidase: NitN amidase from *Nesterenkonia* sp. AN1 (PDB code, 5JQN). (**C**) The representative structures containing structural feature III (SF III) according to the β6–β7 hairpin. The KlNit2 structure is colored in cyan within a black rectangular box. The two structural homologs with different conformations regarding the β6–β7 hairpin are shown in gray. The corresponding β6–β7 hairpin region is shown in red. The two homologous structures indicate the following: HpCN hydrolase: CN hydrolase from Helicobacter pylori (PDB code, 6MG6), PaHT_nitrilase: hyperthermophilic nitrilase from *Pyrococcus abyssi* (PDB code, 3KLC). (**D**) The representative structures containing structural feature IV (SF IV) according to the β11–β12 structure. The KlNit2 structure is colored in cyan within a black rectangular box. The three structural homologs with a different conformation regarding the β11–β13 are shown in gray. The corresponding β6–β7 hairpin region is shown in red. The three homologous structures indicate the following: MtGlu-dep_NAD+ synthetase: glutamine dependent NAD+ synthetase from Mycobacterium tuberculosis (PDB code, 3SZG), SaPCNF_Hydrolase: putative CN family hydrolase from Staphylococcus aureus (PDB code, 3P8K), SspNit6803: Nit6803 from *Synechocystis* sp. (PDB code, 3WUY). (**E**) The representative structures containing structural feature V (SF V) according to the C-terminus. The KlNit2 structure is colored in cyan within a black rectangular box. The six structural homologs with a different conformation regarding the C-terminus are shown in gray. The corresponding C-terminal region is shown in red. The six homologous structures indicate the following: BtGlu-dep_NAD+ synthetase: glutamine dependent NAD+ synthetase from *Burkholderia thailandensis* (PDB code, 4F4H). ChGlu-dep_NAD+ synthetase: glutamine dependent NAD+ synthetase from *Cytophaga hutchinsonii* (PDB code, 3ILV), PaAliphatic amidase: aliphatic amidase from *Pseudomonas aeruginosa* (PDB code, 2UXY), PaHT_nitrilase: hyperthermophilic nitrilase from *Pyrococcus abyssi* (PDB code, 3KLC), XcPNit: putative Nit from *Xanthomonas campestris* (PDB code, 2E11), HsPantetheinase: Vanin-1 from *Homo sapiens* (PDB code, 4CYF). The figures were generated using the graphics program PyMOL (Schrödinger Inc., New York, NY, USA).

Structural feature II contains four types of structures based on the α3 helix between the β3 and β4 strands shown in *Kl*Nit2 (Figure 8B). Except for *Kl*Nit2 and *Sc*Nit2, the other types have no α-helix between the β3 and β4 strands, and the length of the two strands varied. Amidohydrolase, pyrimidine degrading enzyme, two NAD+ synthetases, and several Nit proteins exhibit longer strands than *Kl*Nit2. Formamidase, beta-ureidopropionase, nitrilase, pantetheinase, and three NAD+ synthetases contain strands of similar length as *Kl*Nit2. Shorter strands could be found in two *Nesterenkonia* amidases.

The third feature, the β6–β7 hairpin structure, could be divided into three types (Figure 8C). Most Nit proteins, including *Kl*Nit2, have a clear β6–β7 hairpin. Other types contain no β strands and have one or two turns of α-helix with a loop instead. A protruding loop with a single turn of α-helix could be found in amidohydrolase, pyrimidine degrading enzymes, and NAD+ synthetase. Proteins with a two-turn α-helix include two amidohydrolases, formamidase, nitrilase, pantetheinase, and aliphatic amidase.

We were not able to determine the local conformation between the β11 and β12 strands in *Kl*Nit2 due to the lack of electron density. Regarding this feature, we made three categories, except for a protein that failed to reveal the similar region as observed in *Kl*Nit2 (Figure 8D). The first type showed entirely the loop found in *Sc*Nit3, *Mm*Nit2, *Mycobacterium tuberculosis* glutamine-depending NAD+ synthetase, and pantetheinase. The second type exhibited a hairpin structure with two β strands found in most proteins, such as amidohydrolase, formamidase, aliphatic amidase, and NAD+ synthetase. The last type involved two α-helices connected by loops found in *Synechocystis* sp. Nit6803.

The C-terminal region showed the most variability and could be divided into seven types among the *Kl*Nit2 structural homologs (Figure 8E). The first type contained an α-helix or winding loop after the α6 helix shown in *Kl*Nit2, other Nit proteins, amidohydrolase, pyrimidine degrading enzyme, amidase from *Nesterenkonia*, and several uncharacterized hydrolases. Second, a β-strand instead of the α-helix after the β14 strand was included in three NAD+ synthetases. The third type showed no additional structure after the β14, similar to the one in *Ch*Glu-dependent NAD+ synthetase. The fourth type concerned proteins with a long C-terminal structure including five α-helices after the β14 that could be found in formamidases and aliphatic amidases. The fifth type contained a β-hairpin between the two α-helices shown in *Pyrococcus abyssi* aliphatic amidase and uncharacterized nitrilase-related protein PH0642. The sixth type represented a β-strand after the α-helix shown in Nit4, one of NAD+ synthetases, and putative Nit proteins. The final type is a structure containing one two-turn α-helix with a long loop shown in pantetheinase.

4. Discussion

Glutathione (GSH) is composed of a tripeptide (γ-glutamylcysteinylglycine), which could serve as a potential alternative substrate for various enzymes involved in amino acid and protein metabolism [22]. The γ-glutamyl moiety of the GSH is a structural analog of glutamate that acts as a common amine donor for aminotransferases [23]. Both the cytosolic and mitochondrial aminotransferases in mammals could produce deaminated glutathione (dGSH) by deaminating the amino group of the GSH γ-glutamyl moiety [5]. Potentially harmful compounds, such as dGSH, should be eliminated or transformed into other useful compounds. In that sense, mammalian Nit1 plays an important role in converting dGSH by aminotransferase as a useful metabolite. In addition, Arabidopsis Nit1 enzymes extracted from leaves also have an activity for converting GSH into dGSH in the presence of glyoxylate as an amino acceptor [24]. The mammalian Nit1 protein is a homolog of the yeast Nit2 protein [2].

The overall fold of *Kl*Nit2 mainly exhibits an α–β–β–α architecture, similar to those of homologous structures. However, when we expand our comparison to other structures of the Nit family proteins, five distinct structural features, including an α2 helix, a β3–α3–β4, β6–β7 hairpin, a β11–β12 structure, and a C-terminal region, could be found that might potentially contribute to molecular evolution. Considering the *Sc*Nit2 structure, we found two major distinct regions, where *Kl*Nit2 exhibited a more extended β6–β7 hairpin and disordered α-helices between the β11 and β12 strands due to a crystallographic packing effect (Figure 6). This region might be considered as a framework for a structure-based functional study.

In the *Kl*Nit2 active site, most residues, including the catalytic triads, existed in locations similar to that of *Sc*Nit2. However, the active sites of *Sc*Nit2 and *Mm*Nit2 were larger than that of *Kl*Nit2, which might indicate different substrate specificities [3,18]. Interestingly, the Teng group unexpectedly found an unidentified ligand as an α-keto analog of GSH in the active site of the ScNit2 C169S mutant [18]. They assumed that the molecule could be derived from *E. coli*. Regarding their finding, we also tested whether the *Kl*Nit2 C167A mutant could be complexed with the GSH-like molecule. However, we could not obtain any ligand in the active site of the *Kl*Nit2 C167S mutant. This might be due to the difference in the size of the substrate binding cavity between *Kl*Nit2 and *Sc*Nit2.

In summary, we determined the crystal structure of Nit2 from a fungal species *Kluyveromyces lactis* at a resolution of 2.2 Å. Our extensive comparative analysis of *Kl*Nit2 and its structural homologs revealed five distinct structural features and differences in the size of their active sites. This study could potentially provide new insights into the structural relationships among Nit2 family proteins. Further investigation would be required to reveal the sequence-based molecular structural aspect of their evolution.

Author Contributions: Conceptualization, J.H.C.; methodology, B.-C.J.; investigation, C.J., D.-H.C., H.-S.C., W.-K.K., and J.H.C.; data curation, C.J. and H.J.; writing—original draft preparation, C.J. and J.H.C.; writing—review and editing, C.J. and J.H.C.; visualization, C.J. and J.H.C.; supervision, J.H.C.; and funding acquisition, J.H.C. All authors have read and agreed to the published version of the manuscript.

Funding: This research was supported by Basic Science Research Program through the National Research Foundation of Korea funded by the Ministry of Science and ICT (grant No. NRF-2019R1A2C4069796), and from a cooperation project funded by Korea Institute of Toxicology (KK-2101-01) to J.H.C.

Data Availability Statement: Data is contained within the article.

Acknowledgments: We would like to thank the beamline staff Yeon-Gil Kim and Sung Chul Ha at beamlines 5C and 7A of the Pohang Accelerator Laboratory (Pohang, Korea) for data collection.

Conflicts of Interest: The authors declare no conflict of interest.

References

1. Brenner, C. Catalysis in the nitrilase superfamily. *Curr. Opin. Struct. Biol.* **2002**, *12*, 775–782. [CrossRef]
2. Pace, H.C.; Brenner, C. The nitrilase superfamily: Classification, structure and function. *Genome Biol.* **2001**, *2*, REVIEWS0001. [CrossRef] [PubMed]
3. Barglow, K.T.; Saikatendu, K.S.; Bracey, M.H.; Huey, R.; Morris, G.M.; Olson, A.J.; Stevens, R.C.; Cravatt, B.F. Functional proteomic and structural insights into molecular recognition in the nitrilase family enzymes. *Biochemistry* **2008**, *47*, 13514–13523. [CrossRef] [PubMed]
4. Huebner, K.; Saldivar, J.C.; Sun, J.; Shibata, H.; Druck, T. Hits, Fhits and Nits: Beyond enzymatic function. *Adv. Enzym. Regul.* **2011**, *51*, 208–217. [CrossRef] [PubMed]
5. Peracchi, A.; Veiga-Da-Cunha, M.; Kuhara, T.; Ellens, K.W.; Paczia, N.; Stroobant, V.; Seliga, A.K.; Marlaire, S.; Jaisson, S.; Bommer, G.T.; et al. Nit1 is a metabolite repair enzyme that hydrolyzes deaminated glutathione. *Proc. Natl. Acad. Sci. USA* **2017**, *114*, E3233–E3242. [CrossRef]
6. Maras, B.; Barra, D.; Duprè, S.; Pitari, G. Is pantetheinase the actual identity of mouse and human vanin-1 proteins? *FEBS Lett.* **1999**, *461*, 149–152. [CrossRef]
7. Jaisson, S.; Veiga-da-Cunha, M.; van Schaftingen, E. Molecular identification of omega-amidase, the enzyme that is functionally coupled with glutamine transaminases, as the putative tumor suppressor Nit2. *Biochimie* **2009**, *91*, 1066–1071. [CrossRef]
8. Pace, H.; Hodawadekar, S.; Draganescu, A.; Huang, J.; Bieganowski, P.; Pekarsky, Y.; Croce, C.; Brenner, C. Crystal structure of the worm NitFhit Rosetta Stone protein reveals a Nit tetramer binding two Fhit dimers. *Curr. Biol.* **2000**, *10*, 907–917. [CrossRef]
9. Mittag, S.; Valenta, T.; Weiske, J.; Bloch, L.; Klingel, S.; Gradl, D.; Wetzel, F.; Chen, Y.; Petersen, I.; Basler, K.; et al. A novel role for the tumour suppressor Nitrilase1 modulating the Wnt/beta-catenin signalling pathway. *Cell Discov.* **2016**, *2*, 15039. [CrossRef]
10. Wang, Y.A.; Sun, Y.; le Blanc, J.M.; Solomides, C.; Zhan, T.; Lu, B. Nitrilase 1 modulates lung tumor progression in vitro and in vivo. *Oncotarget* **2016**, *7*, 21381–21392. [CrossRef] [PubMed]
11. Lin, C.-H.; Chung, M.-Y.; Chen, W.-B.; Chien, C.-H. Growth inhibitory effect of the human NIT2 gene and its allelic imbalance in cancers. *FEBS J.* **2007**, *274*, 2946–2956. [CrossRef]
12. Semba, S.; Han, S.-Y.; Qin, H.R.; McCorkell, K.A.; Iliopoulos, D.; Pekarsky, Y.; Druck, T.; Trapasso, F.; Croce, C.M.; Huebner, K. Biological functions of mammalian Nit1, the counterpart of the invertebrate NitFhit Rosetta stone protein, a possible tumor suppressor. *J. Biol. Chem.* **2006**, *281*, 28244–28253. [CrossRef]
13. Zhang, H.; Hou, Y.-J.; Han, S.-Y.; Zhang, E.C.; Huebner, K.; Zhang, J. Mammalian nitrilase 1 homologue Nit1 is a negative regulator in T cells. *Int. Immunol.* **2009**, *21*, 691–703. [CrossRef]
14. Lin, C.; Zhang, J.; Lu, Y.; Li, X.; Zhang, W.; Zhang, W.; Lin, W.; Zheng, L.; Li, X. NIT1 suppresses tumour proliferation by activating the TGFbeta1-Smad2/3 signalling pathway in colorectal cancer. *Cell Death Dis.* **2018**, *9*, 263. [CrossRef]
15. Jin, C.; Jin, H.; Quade, B.; Chang, J.H. Purification, crystallization, and X-ray crystallographic analysis of Nit2 from *Kluyveromyces lactis*. *Biodesign* **2020**, *8*, 20–23. [CrossRef]

16. Adams, P.D.; Afonine, P.V.; Bunkóczi, G.; Chen, V.B.; Davis, I.W.; Echols, N.; Headd, J.J.; Hung, L.-W.; Kapral, G.J.; Grosse-Kunstleve, R.W.; et al. PHENIX: A comprehensive Python-based system for macromolecular structure solution. *Acta Crystallogr. D Biol. Crystallogr.* **2010**, *66 Pt 2*, 213–221. [CrossRef]
17. Emsley, P.; Cowtan, K. Coot: Model-building tools for molecular graphics. *Acta Crystallogr. D Biol. Crystallogr.* **2004**, *60*, 2126–2132. [CrossRef] [PubMed]
18. Liu, H.; Gao, Y.; Zhang, M.; Qiu, X.; Cooper, A.J.L.; Niu, L.; Teng, M. Structures of enzyme-intermediate complexes of yeast Nit2: Insights into its catalytic mechanism and different substrate specificity compared with mammalian Nit2. *Acta Crystallogr. D Biol. Crystallogr.* **2013**, *69 Pt 8*, 1470–1481. [CrossRef]
19. Sakai, N.; Tajika, Y.; Yao, M.; Watanabe, N.; Tanaka, I. Crystal structure of hypothetical protein PH0642 from Pyrococcus horikoshii at 1.6A resolution. *Proteins* **2004**, *57*, 869–873. [CrossRef]
20. Wang, W.-C.; Hsu, W.-H.; Chien, F.-T.; Chen, C.-Y. Crystal structure and site-directed mutagenesis studies of N-carbamoyl-D-amino-acid amidohydrolase from Agrobacterium radiobacter reveals a homotetramer and insight into a catalytic cleft. *J. Mol. Biol.* **2001**, *306*, 251–261. [CrossRef] [PubMed]
21. Holm, L. DALI and the persistence of protein shape. *Protein Sci.* **2020**, *29*, 128–140. [CrossRef] [PubMed]
22. Forman, H.J.; Zhang, H.; Rinna, A. Glutathione: Overview of its protective roles, measurement, and biosynthesis. *Mol. Asp. Med.* **2009**, *30*. [CrossRef] [PubMed]
23. Walker, M.C.; van der Donk, W.A. The many roles of glutamate in metabolism. *J. Ind. Microbiol. Biotechnol.* **2016**, *43*, 419–430. [CrossRef] [PubMed]
24. Niehaus, T.D.; Patterson, J.A.; Alexander, D.C.; Folz, J.S.; Pyc, M.; MacTavish, B.S.; Bruner, S.D.; Mullen, R.T.; Fiehn, O.; Hanson, A.D. The metabolite repair enzyme Nit1 is a dual-targeted amidase that disposes of damaged glutathione in Arabidopsis. *Biochem. J.* **2019**, *476*, 683–697. [CrossRef]

 MDPI

Article

Crystal Structure of an Active Site Mutant Form of IRG1 from *Bacillus subtilis*

Hyun Ho Park

College of Pharmacy, Chung-Ang University, Seoul 06974, Korea; xrayleox@cau.ac.kr

Abstract: Immune-responsive gene1 (IRG1), an enzyme that is overexpressed during immune reactions, catalyzes the production of itaconate from cis-aconitate. Itaconate is a multifunctional immuno-metabolite that displays antibacterial and antiviral activities. The recent resolution of its structure has enabled the mechanism underlying IRG1 function to be speculated on. However, the precise mechanism underlying the enzymatic reaction of IRG1 remains vague owing to the absence of information regarding the structure of the IRG1/substrate or the product complex. In this study, we determined the high-resolution structure of the active site mutant form of IRG1 from *Bacillus subtilis* (bsIRG1_H102A). Structural analysis detected unidentified electron densities around the active site. Structural comparison with the wildtype revealed that H102 was critical for the precise location of the side chain of residues around active site of IRG1. Finally, the activity of bsIRG1 was extremely low compared with that of mammalian IRG1. The current structural study will expectedly help understand the working mechanism of IRG1.

Keywords: crystal structure; immunometabolite; itaconate; immune-responsive gene 1

Citation: Park, H.H. Crystal Structure of an Active Site Mutant Form of IRG1 from *Bacillus subtilis*. *Crystals* **2021**, *11*, 350. https://doi.org/10.3390/cryst11040350

Academic Editor: Kyeong Kyu Kim

Received: 11 March 2021
Accepted: 26 March 2021
Published: 29 March 2021

1. Introduction

Itaconate is an unsaturated dicarboxylic acid, which functions as an intermediate metabolite in the tricarboxylic acid (TCA) cycle [1]. Owing to the wide usage of itaconate in industrial applications, such as the production of various resins and bioactive compounds, itaconate has become the subject of intensive material science-based research [2,3]. However, in biology, itaconate is known for the role it plays in regulating innate immunity as well as for its antibacterial and antiviral activities [4–6]. Antimicrobial activity of itaconate was initially reported in fungi [7]. *Aspergillus terreus,* a representative fungus, produces high levels of itaconate that act against *Pseudomonas indigofera* and *Mycobacterium tuberculosis* by directly inhibiting the activity of housekeeping enzymes, such as isocitrate lyase (ICL) and fructose-6-phosphate 2-kinase [8–10]. The antimicrobial activity of itaconate in the mammalian system has also been revealed recently [4–6]. Itaconate eliminates infections caused by bacteria, such as *Salmonella enterica* and *M. tuberculosis*, by targeting ICL [4]. Itaconate reportedly inhibits succinate dehydrogenase activity during Zika virus infections [5].

Immune responsive gene 1 (IRG1), also known as cis-aconitate decarboxylase (CAD), is a ~55 kDa enzyme that catalyzes the decarboxylation of cis-aconitate to generate itaconate [11,12]. Recent mammalian system studies indicate that IRG1, which is overexpressed in macrophages in response to pathogen-associated molecular patterns (PAMPs) such as lipopolysaccharides (LPS), catalyzes the overproduction of itaconate against pathogenic infections [4,13–15]. Since excessive production of itaconate caused by abnormal overexpression of IRG1 induces gout [16], chronic arthritis [17], and tumor progression in mouse models [18], targeting IRG1 may lead to therapeutic intervention in various human diseases.

Due to increased interest in the role of itaconate as well as its production, the mechanism underlying IRG1-mediated itaconate production, using cis-aconitate as a substrate, is being intensively researched biochemically as well as structurally [19–21]. Structural studies of mammalian IRG1 have helped determine the dynamic open-closed structure of IRG1 as well as its putative active site [19,20], while the results of structural studies conducted on IRG1 from *B. subtilis* have led to a tentative functional mechanism being proposed for IRG1 [21]. Despite recent structural and enzymatic studies, the precise catalytic mechanism of IRG1 in the innate immunity process has remained elusive. In this study, we resolved the high-resolution structure of active site mutant form of IRG1 from *B. subtilis* (bsIRG1_H102A). Comparative structural analyses, accompanied by an enzymatic activity assay, were used to determine features of bsIRG1_H102A. Our current investigation of the active site mutant form of bsIRG1 may provide helpful information regarding the substrate binding process of IRG1 and the functional mechanism underlying this process.

2. Materials and Methods

2.1. Site-Directed Mutagenesis

A quickchange mutagenesis kit (Stratagene) was used for site-directed mutagenesis. The expression plasmid containing a full-length *IRG1* gene from *B. subtilis* (corresponding to amino acids 1–445) was used for the template. Induced mutagenesis of H102A was then confirmed via sequencing.

2.2. Protein Expression and Purification

The plasmid encoding bsIRG1_H102A was transformed into *Escherichia coli* BL21 (DE3) cells. A single colony was picked and cultured in 5 mL lysogeny broth (LB) medium containing 50 µg/mL kanamycin overnight at 37 °C. Cells from the small culture were then transferred and cultured in 1 L of medium until optical density (OD) of approximately 0.65 was reached at 600 nm, at which point 0.3 mM isopropyl β-D-1-thiogalactopyranoside (IPTG) was added to the medium. Next, the cells were further cultured for 22 h at 20 °C. Cells, collected after harvesting via centrifugation at 20 °C, were suspended with 40 mL of lysis buffer (20 mM Tris-HCl [pH 8.0], 500 mM NaCl, and 25 mM imidazole) containing a serine protease inhibitor (phenylmethanesulfonyl fluoride; Sigma-Aldrich, St. Louis, MO, USA). After the cells were lysed via sonication, cell debris was removed by centrifugation at 10,000× g for 30 min at 4 °C. Then, the supernatant was collected and mixed with nickel nitrilotriacetic acid (Ni-NTA) resin (Qiagen, Hilden, Germany) using gentle agitation, for 4 h at 4 °C. The resulting mixture was applied to a gravity-flow column and washed with 50 mL of washing buffer (20 mM Tris-HCl [pH 8.0], 500 mM NaCl, and 60 mM imidazole). Next, a total of 2 mL of elution buffer (20 mM Tris-HCl [pH 7.9], 500 mM NaCl, and 250 mM imidazole) was applied to the column to elute the bound target protein. The eluted target protein was concentrated to 30 mg/mL and sequentially subjected to size-exclusion chromatography, which was conducted via an ÄKTA explorer system (GE Healthcare, Chicago, IL, USA) equipped with a Superdex 200 Increase 10/300 GL 24 mL column (GE Healthcare) pre-equilibrated with SEC buffer (20 mM Tris-HCl [pH 8.0] and 150 mM NaCl). Peak fractions were pooled, concentrated to 8.5 mg/mL, flash-frozen in liquid N_2, and stored at −80°C until further use. Mouse and human IRG1 proteins, used for the activity study, were prepared using the same method used to purify bsIRG1 protein [21]. The accession numbers (NCBI reference sequences) for the sequence of bsIRG1, mouseIRG1, and humanIRG1 are ARW33836, NP_032418, and NP_001245335, respectively.

2.3. Crystallization and Data Collection

Crystallization of bsIRG1_H102A was conducted via the same method used for wildtype bsIRG1, and crystals were obtained under conditions that were similar to those used to produce wildtype crystals [20]. Briefly, 1 µL of protein solution was mixed with an equal volume of reservoir solution containing 1.4 M ammonium sulfate, 0.1 M CAPS pH 10.2, and 0.2 M lithium sulfate and the droplets were allowed to equilibrate against 300 µL of mother liquor using the hanging drop vapor diffusion method at 20 °C. For purpose of data collection, crystals were soaked in a cryoprotectant solution consisting of mother liquor supplemented with 40% (v/v) glycerol. These crystals were flash-cooled in a N_2 stream at −178 °C. X-ray diffraction data were collected at the Pohang Accelerator Laboratory with the 5C beamline (Pohang, Republic of Korea). The diffraction data were indexed, integrated, and scaled using the HKL-2000 program [22].

2.4. Structure Determination and Analysis

The molecular replacement (MR) phasing method was used for initial phasing determination. A PHASER [23] in PHENIX package was employed for MR. The previously solved wild-type structure of bsIRG1 (PDB ID: 7BRA) was used as the search model [20]. The initial model was built automatically with AutoBuild in PHENIX [24]. Further model building and refinement were performed using Coot [25] and Phenix.refine in PHENIX package [24]. Model quality was validated using MolProbity [26]. All structural figures were produced via the PyMOL program [27].

2.5. Decarboxylation Activity Test and HPLC Analysis

To conduct enzymatic activity tests, a final 100 µL reaction solution was prepared in 25 µM HEPES buffer pH 7.1 supplied with 0.5 µM of each IRG1 enzyme and 1.5 mM cis-aconitate as substrate. The reaction, which proceeded for 1 h at 30°C, was stopped by adding 400 µL methanol [28]. Next, the reaction mixture was analyzed using Agilent High-Pressure Liquid Chromatography (HPLC) equipped with an ACC-3000 autosampler, Mightysil RP-18 GP reverse-phase C18 column (150 V, 4.6 mm, Japan) and a DAD-3000 diode array detector. The mobile phase consisted of water mixed with 0.1% trifluoroacetic acid as solution A and HPLC-grade acetonitrile as solution B. Substrates and their products were detected via a UV detector at 210 nm. All HPLC experiments were performed at 30°C. The flow rate used to separate the sample was set to 0.9 mL/min, and solution A was maintained at 95% for 40 min for analysis.

2.6. Structural Data Accession Number

Coordinate and structural factors were deposited with the Protein Data Bank under PDB ID: 7E9D.

3. Results

3.1. Preliminary X-ray Crystallographic Studies of bsIRG1_H102A

IRG1 converts cis-aconitate to itaconate, an important immuno-metabolite that acts against pathogens (Figure 1a). Previous structural and enzymatic studies have indicated that seven residues are present around the active site. These residues, namely D93 (D91 in bsIRG1), T97 (T95 in bsIRG1), H103 (H102 in bsIRG1), H159 (H151 in bsIRG1), K207 (K200 in bsIRG1), K272 (K266 in bsIRG1), and Y318 (A310 in bsIRG1), play a critical role in mammalian IRG1 activity [19,20]. To understand the precise enzymatic mechanism of bsIRG1, together with the function of H102, we attempted to characterize and solve the structure of bsIRG1_H102A.

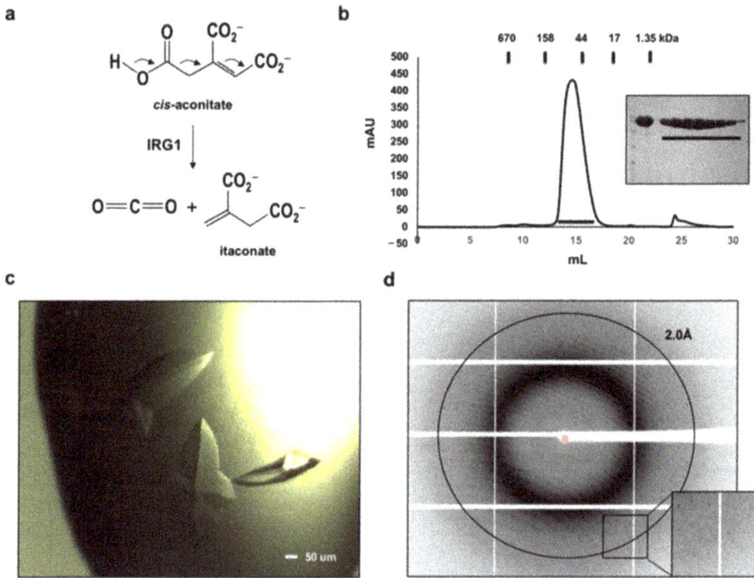

Figure 1. Crystallization and X-ray crystallographic studies of bsIRG1_H102A. (**a**) Schematic of the enzymatic reaction of IRG1 that produce itaconate using cis-aconitate as a substrate. (**b**) Profile of FPLC used for purification of bsIRG1_H102A. SDS-PAGE was provided at the right side of peak. Black bar indicates the fractions from FPLC loaded onto SDS-PAGE. The position of size standard markers is also provided above the FPLC profile. (**c**) The crystal of bsIRG1_H102A. A scale bar is shown at the bottom right-hand corner of the crystal image. (**d**) Diffraction image. Diffraction spots around 2.0 Å were magnified.

In preparation for structural and biochemical studies, the active site mutant form of bsIRG1 (bsIRG1_H102A) was purified via a quick two step chromatographic procedure. His-tag affinity chromatography followed by size-exclusion chromatography produced a ~95% pure bsIRG1_H102A protein with no contaminating bands, as was observed upon analysis via SDS-PAGE (Figure 1b). The monomeric molecular weight of bsIRG1_H102A including the N-terminal His-tag was estimated to be 47.5 kDa, and its size-exclusion chromatography elution peak suggested that, similar to the wildtype, it exists as a dimer in solution, (Figure 1b). This indicated that introducing the mutant to alanine at H102 did not affect the stoichiometry of bsIRG1 in solution.

The purified protein was crystallized in one day and the crystals were grown to a maximum size of $0.5 \times 0.1 \times 0.1$ mm^3 (Figure 1c). The crystals were diffracted to 1.89 Å at the synchrotron (Figure 1d). The crystals belonged to space group $P2_12_12_1$ with unit-cell parameters of a = 58.91, b = 110.77, c = 168.56 Å. Diffraction statistics are shown (Table 1).

Table 1. Crystallographic statistics.

Data Collection	bsIRG1_H102A
X-ray source	Synchrotron (PAL 5C)
Detector	Eiger 9M
Wavelength	0.97950
Space group	$P2_12_12_1$
Cell dimensions	
a, b, c	58.91 Å, 110.77 Å, 168.56 Å
α, β, γ	90°, 90°, 90°
† Resolution	50.00–1.89 (1.93–1.89) Å
Wilson B-factor	21.38 $Å^2$
† No. of unique reflections overall	85,234 (5,322)
† Rsym	0.064 (0.754)
† I/σI	18.7 (2.1)
† Completeness	95.6% (91.3%)
† Redundancy	9.8 (8.9)
CC1/2	0.99 (0.79)
Refinement	
† Resolution	40.35–1.89 (1.91–1.89) Å
No. of reflections used (completeness)	85,146 (95.5%)
No. of non-H protein atoms	6,688
No. of water molecules	763
No. of ions	0
† Rwork	17.21% (24.24%)
† Rfree	20.21% (26.57%)
Average B-factors	
Protein	22.9 $Å^2$
Water molecules	34.4 $Å^2$
r.m.s. deviations	
Bond lengths	0.006 Å
Bond angles	0.774°
Ramachandran Plot	
Ramachandran outliers	0.00%
Ramachandran favored	99.32%
Ramachandran allowed	0.68%
Rotamer outliers	0.00%
Clash score	2.25

† The highest resolution shell is presented in parenthesis. r.m.s, root mean square.

3.2. Overall Structure of H102A Mutant Form of bsIRG1

The structure of bsIRG1_H102A was solved using the molecular replacement phasing method with wildtype structure as the search model. The structure was refined to R_{work} = 17.21% and R_{free} = 20.21%. The refinement statistics are summarized in Table 1. The asymmetric unit comprised two molecules, with the final model encompassing residues 5–445 for both molecules (Figure 2a). The structures of the two molecules in the asymmetric unit were nearly identical, with a root mean square deviation (RMSD) of 0.46 Å. Phylogenetic analysis of homologous IRG1 sequences from various species using the ConSurf server [29] indicated that the residues involved in the formation of the putative active site, including H102, were highly conserved as expected (Figure 2b).

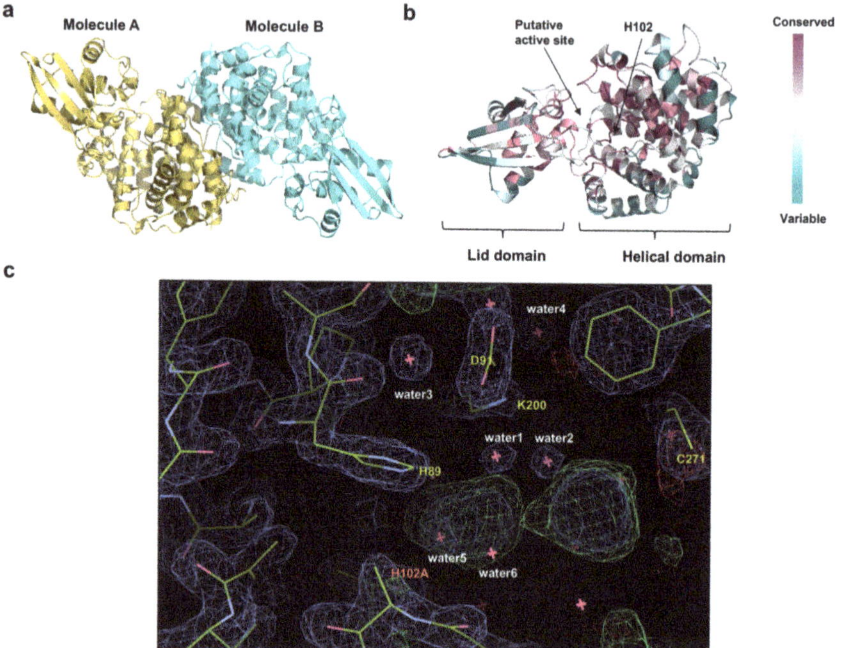

Figure 2. Structure of bsIRG1_H102A and the active site around H102. (**a**) Crystal structure of bsIRG1_H102A. Two molecules in an asymmetric unit, molecules A and B, are presented via a cartoon model. (**b**) Cartoon representation of bsIRG1 with colors indicating the degree of amino acid sequence conservation, as analyzed by Consurf. Two distinct IRG1 domains, the lid and helical domains, are indicated. The positions of the putative active site and H102 site are indicated by black arrows (**c**) 2fo-fc map (blue mesh) and fo-fc map (green mesh) contoured at the 1-σ level for 2fofc map and 2.5-σ level around the mutated residue, H102A.

The first structural analysis that was performed using this high-resolution structure led to the identification of the mutation site, H102. Absolute absence of the electron density corresponding to the histidine side chain at the H102 site indicated that the H102 residue had successfully mutated to alanine via mutagenesis (Figure 2c). Because the H102 residue in the putative active site of IRG1 is one of the residues critical for IRG1 activity, our structural analysis of bsIRG1_H102A focused on the region around the H102 site, where we observed two large unidentified bulb-like densities, which had a much higher density than that of water but were not structurally fit to accommodate either the substrate or product (Figure 2c).

3.3. Comparison of the Structure of bsIRG1_H102A with the Structure of Wildtype bsIRG1

Structural comparison between the wildtype and mutant via superimposition analysis indicated that the structures were nearly identical, exhibiting a RMSD of 0.483 Å and TM-score of 0.923 (Figure 3a). Despite such structural similarities, structural comparison between bsIRG1_H102A and the wildtype, with particular reference to the putative active site around the H102 residue, revealed that the location of the side chains of several residues, including those of H89, R100, and K200, were altered due to H102A mutagenesis (Figure 3b). Loss of the bulky side chain of histidine due to mutagenesis allowed the ring structure of H89 to rotate causing the location of the side chain of K200 to move slightly (Figure 3b). The most dramatic change in the structure of the side chain was detected at the R100 residue. The temperature factors of residue R100 in the wildtype and mutant protein are 32 Å^2 and 35 Å^2, respectively. Introduction of the mutation caused the side chain of this residue to locate away from H102 (Figure 3b).

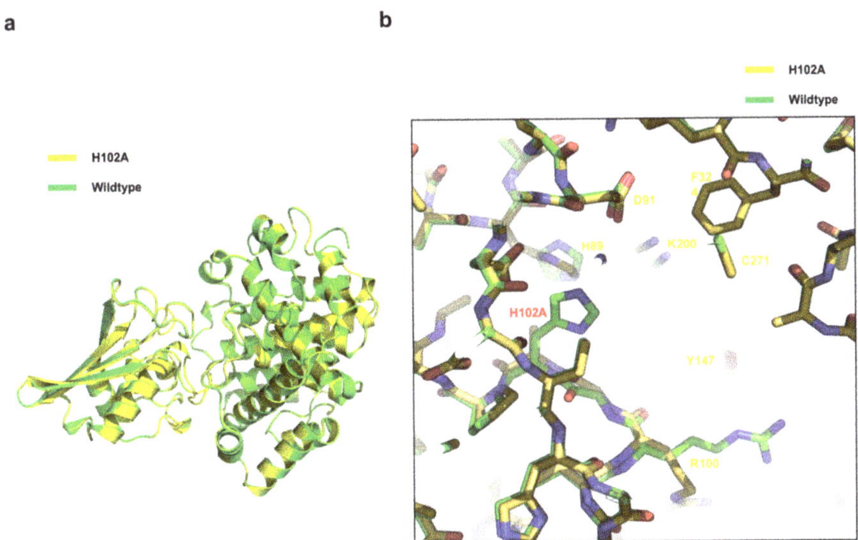

Figure 3. Structural comparison of bsIRG1_H102A with wildtype bsIRG1: (**a**) superimposition of bsIRG1_H102A (yellow color) on wildtype bsIRG1 (green color); and (**b**) magnified view of the active site around the H102 residue.

3.4. Activity Comparison of bsIRG1 with Mammalian IRG1

To confirm the functional importance of the H102 residue in bsIRG1 (H103 residue in mammalian IRG1), the activities of both wildtype bsIRG1 and bsIRG1_H102A were tested using high-pressure liquid chromatography (HPLC). Before the activity test, the standard elution positions of cis-aconitate and itaconate were determined via HPLC (Figure 4a,b). For the enzymatic activity test, the reaction mixture, including cis-aconitate and wildtype bsIRG1 or bsIRG1_H102A, was loaded onto the HPLC apparatus, following which itaconate production was monitored by analyzing the eluted position of the newly produced compound. The HPLC profile showed that only a minute amount of itaconate was produced by wildtype bsIRG1 (Figure 4c), whereas no itaconate at all was produced by the bsIRG1_H102A mutant (Figure 4d). Because the activity of bsIRG1 in this experiment was seen to be low, we compared the activity of bsIRG1 with mammalian IRG1 (IRG1 from mouse and human). This experiment showed that similar amounts of itaconate (although a much larger amount when compared with the peak corresponding to wildtype bsIRG1), were produced by both mouse IRG1 and human IRG (Figure 4e,f), indicating that mammalian IRG1 displayed stronger cis-aconitate decarboxylase activity compared with that of bsIRG1.

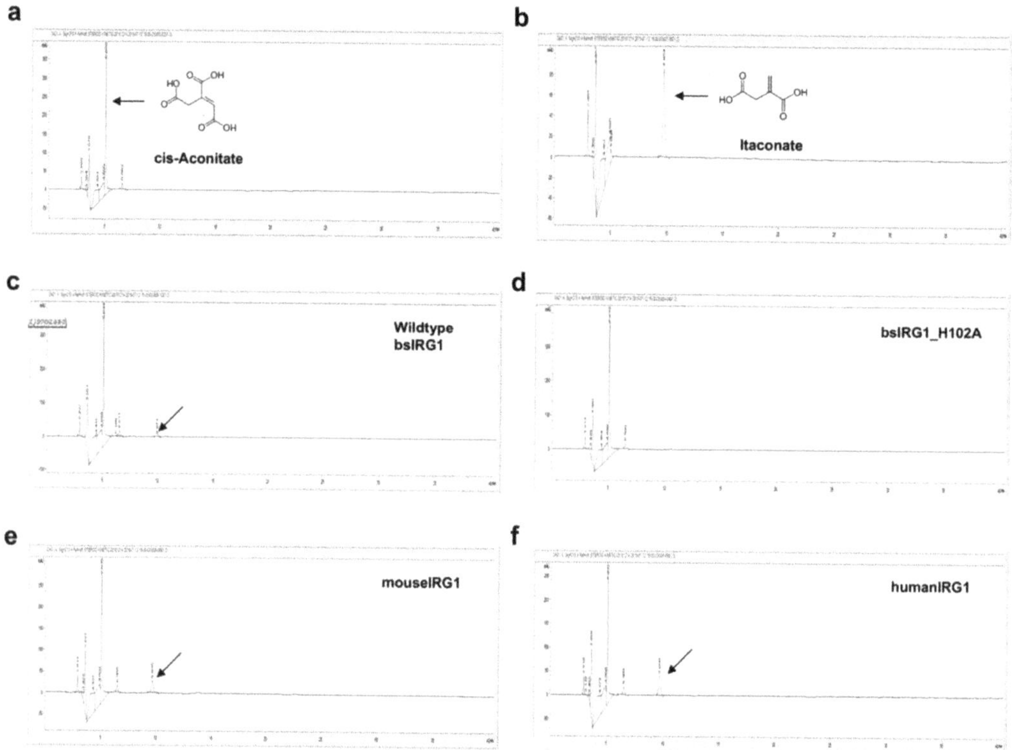

Figure 4. Comparative analysis of bsIRG1 and mammalian IRG1 activities. (**a**,**b**) HPLC profiles. The eluted standard positions of cis-aconitate (**a**) and itaconate (**b**) are indicated by black arrows. (**c**) HPLC profile showing wildtype bsIRG1 activity. The peak produced by the enzymatic reaction is indicated by a black arrow. (**d**) HPLC profile showing bsIRG1_H102A activity. No peak was observed for itaconate. (**e**,**f**) HPLC profiles showing the activities of mouseIRG1 and humanIRG1. The peaks produced by the enzymatic reaction are indicated by a black arrow.

4. Discussion

To understand the precise mechanism by which IRG1 catalyzes the production of itaconate during microbial infection, we solved the structure of the active site mutant variant of IRG1 found in *B. subtilis* (bsIRG1_H102A). We investigated the manner in which the structure of this variant H102 residue, critical for IRG1 activity, is mutated to alanine. Previous structural and modeling studies have suggested that H102 may be the key residue, because it provides a general base for electron transfer during the decarboxylation process [20].

Biochemical analysis of bsIRG1_H102A variant showed that H102A mutation did not affect the formation of a functional IRG1 dimer, while phylogenic sequence analysis indicated that residues around the putative active site, including H102, were well conserved throughout different species (Figure 2b). During the structural analysis of bsIRG1_H102A, we found two bulb-like electron densities around the active site (near H102) (Figure 2c). These densities, which were much larger than that of water, did not fit either the substrate or the product. Because we did not add any substrate or product during structural determination, these densities may have been produced by ions or unidentified molecules that are present in cells used for protein expression. Since these unidentified densities were located within the pocket of the putative active site, further investigations may be required to identify these densities.

Although the overall structure of the bsIRG1_H102A variant is almost identical with that of wildtype bsIRG1, structure comparison by superimposition analysis indicated that the location of the side chains of H89, R100, and K200, which are critical for the activity of IRG1, were altered by introduction of the mutation. For example, the histidine ring of H89 rotated 90° and the location of the side chain of K200 was slightly moved. In the case of R100, the side chain was kinked to 90° and localized far away from the active site. This observation indicated that H102 may be crucial for accurate formation of the active site, which is critical for the proper activity of IRG1, although H102 has been suggested as one of the key general bases that can transfer electron during the decarboxylation reaction [20].

Interestingly, activity analysis of bsIRG1 indicated that it exhibits extremely low activity compared with mammalian IRG1. Such low bsIRG1 activity may be explained by the failure to find specific reaction conditions. Beside this, our previous finding, which showed that the active site of bsIRG1 was highly similar to the active site of IDS epimerase, may explain the limited nature of bsIRG1 activity [20]. H99 and Y145 in the active site of IDS epimerase are known to be critical for the epimerization reaction [30]. During the previous study, we observed that these two residues were conserved only in bsIRG1, and not in mammalian IRG1 [20]. In addition to this similarity between bsIRG1 and epimerase, differences between the active sites of bsIRG1 and mammalian IRG1 have been also observed. Two residues, H277 and Y318 in human IRG1 (Y319 in mouse IRG1), which were previously identified as critical for IRG1 activity [19], are not conserved in bsIRG1. This structural analysis showed that the active site composition of bsIRG1 is more similar to that of epimerase than to that of mammalian IRG1. Further activity analysis of bsIRG1 may have to be performed to elucidate the unknown causes leading to low IRG1 activity and the relationship between decarboxylase activity and epimerase activity of bsIRG1.

Funding: This study was supported by the National Research Foundation (NRF) of Korea, which is funded by the Ministry of Science, ICT (Information and Communication Technology), and Future Planning (MSIP) (2017M3A9D8062960 and 2018R1A4A1023822) of the Korean Government.

Data Availability Statement: Not applicable.

Acknowledgments: The author thanks members of the Park laboratory for helpful discussions.

Conflicts of Interest: The author declares no conflict of interest.

References

1. O'Neill, L.A.J.; Artyomov, M.N. Itaconate: The poster child of metabolic reprogramming in macrophage function. *Nat. Rev. Immunol.* **2019**, *19*, 273–281. [CrossRef] [PubMed]
2. Kurian, J.V. A new polymer platform for the future—Sorona (R) from corn derived 1,3-propanediol. *J. Polym. Environ.* **2005**, *13*, 159–167. [CrossRef]
3. Cordes, T.; Michelucci, A.; Hiller, K. Itaconic Acid: The Surprising Role of an Industrial Compound as a Mammalian Antimicrobial Metabolite. *Annu. Rev. Nutr.* **2015**, *35*, 451–473. [CrossRef]
4. Michelucci, A.; Cordes, T.; Ghelfi, J.; Pailot, A.; Reiling, N.; Goldmann, O.; Binz, T.; Wegner, A.; Tallam, A.; Rausell, A.; et al. Immune-responsive gene 1 protein links metabolism to immunity by catalyzing itaconic acid production. *Proc. Natl. Acad. Sci. USA* **2013**, *110*, 7820–7825. [CrossRef]
5. Daniels, B.P.; Kofman, S.B.; Smith, J.R.; Norris, G.T.; Snyder, A.G.; Kolb, J.P.; Gao, X.; Locasale, J.W.; Martinez, J.; Gale, M.; et al. The Nucleotide Sensor ZBP1 and Kinase RIPK3 Induce the Enzyme IRG1 to Promote an Antiviral Metabolic State in Neurons. *Immunity* **2019**, *50*, 64–76.e4. [CrossRef]
6. Nair, S.; Huynh, J.P.; Lampropoulou, V.; Loginicheva, E.; Esaulova, E.; Gounder, A.P.; Boon, A.C.M.; Schwarzkopf, E.A.; Bradstreet, T.R.; Edelson, B.T.; et al. Irg1 expression in myeloid cells prevents immunopathology during M-tuberculosis infection. *J. Exp. Med.* **2018**, *215*, 1035–1045. [CrossRef]
7. Bentley, R.; Thiessen, C.P. Biosynthesis of itaconic acid in Aspergillus terreus. III. The properties and reaction mechanism of cis-aconitic acid decarboxylase. *J. Biol. Chem.* **1957**, *226*, 703–720. [CrossRef]
8. Khan, F.R.; Mcfadden, B.A. Enzyme Profiles in Seedling Development and the Effect of Itaconate, an Isocitrate Lyase-Directed Reagent. *Plant Physiol.* **1979**, *64*, 228–231. [CrossRef] [PubMed]
9. Sakai, A.; Kusumoto, A.; Kiso, Y.; Furuya, E. Itaconate reduces visceral fat by inhibiting fructose 2,6-bisphosphate synthesis in rat liver. *Nutrition* **2004**, *20*, 997–1002. [CrossRef]

10. Li, A.; Pfelzer, N.; Zuijderwijk, R.; Punt, P. Enhanced itaconic acid production in Aspergillus niger using genetic modification and medium optimization. *BMC. Biotechnol.* **2012**, *12*, 57. [CrossRef] [PubMed]
11. Dwiarti, L.; Yamane, K.; Yamatani, H.; Kahar, P.; Okabe, M. Purification and characterization of cis-aconitic acid decarboxylase from Aspergillus terreus TN484-M1. *J. Biosci. Bioeng.* **2002**, *94*, 29–33. [CrossRef]
12. Kanamasa, S.; Dwiarti, L.; Okabe, M.; Park, E.Y. Cloning and functional characterization of the cis-aconitic acid decarboxylase (CAD) gene from Aspergillus terreus. *Appl. Microbiol. Biot.* **2008**, *80*, 223–229. [CrossRef] [PubMed]
13. Mills, E.L.; Ryan, D.G.; Prag, H.A.; Dikovskaya, D.; Menon, D.; Zaslona, Z.; Jedrychowski, M.P.; Costa, A.S.H.; Higgins, M.; Hams, E.; et al. Itaconate is an anti-inflammatory metabolite that activates Nrf2 via alkylation of KEAP1. *Nature* **2018**, *556*, 113–117. [CrossRef]
14. Basler, T.; Jeckstadt, S.; Valentin-Weigand, P.; Goethe, R. Mycobacterium paratuberculosis, Mycobacterium smegmatis, and lipopolysaccharide induce different transcriptional and post-transcriptional regulation of the IRG1 gene in murine macrophages. *J. Leukoc. Biol.* **2006**, *79*, 628–638. [CrossRef]
15. Strelko, C.L.; Lu, W.Y.; Dufort, F.J.; Seyfried, T.N.; Chiles, T.C.; Rabinowitz, J.D.; Roberts, M.F. Itaconic Acid Is a Mammalian Metabolite Induced during Macrophage Activation. *J. Am. Chem. Soc.* **2011**, *133*, 16386–16389. [CrossRef]
16. Pessler, F.; Mayer, C.T.; Jung, S.M.; Behrens, E.M.; Dai, L.; Menetski, J.P.; Schumacher, H.R. Identification of novel monosodium urate crystal regulated mRNAs by transcript profiling of dissected murine air pouch membranes. *Arthritis Res. Ther.* **2008**, *10*, R64. [CrossRef] [PubMed]
17. Michopoulos, F.; Karagianni, N.; Whalley, N.M.; Firth, M.A.; Nikolaou, C.; Wilson, I.D.; Critchlow, S.E.; Kollias, G.; Theodoridis, G.A. Targeted Metabolic Profiling of the Tg197 Mouse Model Reveals Itaconic Acid as a Marker of Rheumatoid Arthritis. *J. Proteome Res.* **2016**, *15*, 4579–4590. [CrossRef] [PubMed]
18. Weiss, J.M.; Davies, L.C.; Karwan, M.; Ileva, L.; Ozaki, M.K.; Cheng, R.Y.; Ridnour, L.A.; Annunziata, C.M.; Wink, D.A.; McVicar, D.W. Itaconic acid mediates crosstalk between macrophage metabolism and peritoneal tumors. *J. Clin. Investig.* **2018**, *128*, 3794–3805. [CrossRef]
19. Chen, F.F.; Lukat, P.; Iqbal, A.A.; Saile, K.; Kaever, V.; van den Heuvel, J.; Blankenfeldt, W.; Bussow, K.; Pessler, F. Crystal structure of cis-aconitate decarboxylase reveals the impact of naturally occurring human mutations on itaconate synthesis. *Proc. Natl. Acad. Sci. USA* **2019**, *116*, 20644–20654. [CrossRef]
20. Chun, H.L.; Lee, S.Y.; Lee, S.H.; Lee, C.S.; Park, H.H. Enzymatic reaction mechanism of cis-aconitate decarboxylase based on the crystal structure of IRG1 from Bacillus subtilis. *Sci. Rep.* **2020**, *10*, 11305. [CrossRef] [PubMed]
21. Chun, H.L.; Lee, S.Y.; Kim, K.H.; Lee, C.S.; Oh, T.J.; Park, H.H. The crystal structure of mouse IRG1 suggests that cis-aconitate decarboxylase has an open and closed conformation. *PLoS ONE* **2020**, *15*, e0242383. [CrossRef]
22. Otwinowski, Z.; Minor, W. Processing of X-ray diffraction data collected in oscillation mode. *Methods Enzymol.* **1997**, *276*, 307–326. [PubMed]
23. McCoy, A.J. Solving structures of protein complexes by molecular replacement with Phaser. *Acta Crystallogr. D Biol. Crystallogr.* **2007**, *63*, 32–41. [CrossRef] [PubMed]
24. Adams, P.D.; Afonine, P.V.; Bunkoczi, G.; Chen, V.B.; Davis, I.W.; Echols, N.; Headd, J.J.; Hung, L.W.; Kapral, G.J.; Grosse-Kunstleve, R.W.; et al. PHENIX: A comprehensive Python-based system for macromolecular structure solution. *Acta Crystallogr. D Biol. Crystallogr.* **2010**, *66*, 213–221. [CrossRef]
25. Emsley, P.; Cowtan, K. Coot: Model-building tools for molecular graphics. *Acta Crystallogr. D Biol. Crystallogr.* **2004**, *60*, 2126–2132. [CrossRef] [PubMed]
26. Chen, V.B.; Arendall, W.B., III; Headd, J.J.; Keedy, D.A.; Immormino, R.M.; Kapral, G.J.; Murray, L.W.; Richardson, J.S.; Richardson, D.C. MolProbity: All-atom structure validation for macromolecular crystallography. *Acta Crystallogr. D Biol. Crystallogr.* **2010**, *66*, 12–21. [CrossRef]
27. DeLano, W.L.; Lam, J.W. PyMOL: A communications tool for computational models. *Abstr. Pap. Am. Chem. Soc.* **2005**, *230*, U1371–U1372.
28. Vuoristo, K.S.; Mars, A.E.; van Loon, S.; Orsi, E.; Eggink, G.; Sanders, J.P.; Weusthuis, R.A. Heterologous expression of Mus musculus immunoresponsive gene 1 (irg1) in Escherichia coli results in itaconate production. *Front Microbiol.* **2015**, *6*, 849. [CrossRef] [PubMed]
29. Ashkenazy, H.; Erez, E.; Martz, E.; Pupko, T.; Ben-Tal, N. ConSurf 2010: Calculating evolutionary conservation in sequence and structure of proteins and nucleic acids. *Nucleic Acids Res.* **2010**, *38*, W529–W533. [CrossRef]
30. Lohkamp, B.; Bauerle, B.; Rieger, P.G.; Schneider, G. Three-dimensional structure of iminodisuccinate epimerase defines the fold of the MmgE/PrpD protein family. *J. Mol. Biol.* **2006**, *362*, 555–566. [CrossRef] [PubMed]

Communication

Purification and Crystallographic Analysis of a Novel Cold-Active Esterase (*Ha*Est1) from *Halocynthiibacter arcticus*

Sangeun Jeon [1,†], Jisub Hwang [2,3,†], Wanki Yoo [1,4], Joo Won Chang [1], Hackwon Do [2], Han-Woo Kim [2,3], Kyeong Kyu Kim [4], Jun Hyuck Lee [2,3,*] and T. Doohun Kim [1,*]

1 Department of Chemistry, Graduate School of General Studies, Sookmyung Women's University, Seoul 04310, Korea; sangeun94@sookmyung.ac.kr (S.J.); vlqkshqk61@outlook.kr (W.Y.); cjw1423@sookmyung.ac.kr (J.W.C.)
2 Research Unit of Cryogenic Novel Material, Korea Polar Research Institute, Incheon 21990, Korea; hjsub9696@kopri.re.kr (J.H.); hackwondo@kopri.re.kr (H.D.); hwkim@kopri.re.kr (H.-W.K.)
3 Department of Polar Sciences, University of Science and Technology, Incheon 21990, Korea
4 Department of Molecular Cell Biology, School of Medicine, Sungkyunkwan University, Suwon 16419, Korea; kyeongkyu@skku.edu
* Correspondence: junhyucklee@kopri.re.kr (J.H.L.); doohunkim@sookmyung.ac.kr (T.D.K.)
† These authors contributed equally to this work.

Abstract: This report deals with the purification, characterization, and a preliminary crystallographic study of a novel cold-active esterase (*Ha*Est1) from *Halocynthiibacter arcticus*. Primary sequence analysis reveals that *Ha*Est1 has a catalytic serine in G-x-S-x-G motif. The recombinant *Ha*Est1 was cloned, expressed, and purified. SDS-PAGE and zymographic analysis were carried out to characterize the properties of *Ha*Est1. A single crystal of *Ha*Est1 was obtained in a solution containing 10% (w/v) PEG 8000/8% ethylene glycol, 0.1 M Hepes-NaOH, pH 7.5. Diffraction data were collected to 2.10 Å resolution with $P2_1$ space group. The final R_{merge} and $R_{p.i.m}$ values were 7.6% and 3.5% for 50–2.10 Å resolution. The unit cell parameters were a = 35.69 Å, b = 91.21 Å, c = 79.15 Å, and β = 96.9°.

Keywords: esterase; enzyme assay; crystallization; diffraction

Citation: Jeon, S.; Hwang, J.; Yoo, W.; Chang, J.W.; Do, H.; Kim, H.-W.; Kim, K.K.; Lee, J.H.; Kim, T.D. Purification and Crystallographic Analysis of a Novel Cold-Active Esterase (*Ha*Est1) from *Halocynthiibacter arcticus*. *Crystals* **2021**, *11*, 170. https://doi.org/10.3390/cryst11020170

Academic Editor: Borislav Angelov
Received: 22 January 2021
Accepted: 7 February 2021
Published: 8 February 2021

1. Introduction

Microbial esterase catalyzes the reaction of formation and hydrolysis of chemical bonds between hydroxyl and carboxylic acid groups, which could be used in the preparation of many biological products such as foods, flavors, cosmetics, drugs, and agrochemicals [1–3]. In addition, they are employed in the degradation of carbamates, pesticides, polymer based plastics, and industrial wastes. These enzymes share the characteristic α/β hydrolase fold, conserved catalytic triad of Ser-His-Asp/Glu, catalytic strategies, substrate specificities, and a lack of cofactors. It has been shown that esterases perform catalytic reactions via a nucleophilic attack on the substrates [1,4–6].

Cold-active enzymes show relatively high activity at low temperatures compared to their mesophilic or thermophilic proteins [7,8]. Due to high demands, there have been a lot of studies to characterize cold-active enzymes [9]. To date, cold-active esterases were identified from *Paenibacillus* sp. [10], *Pseudomonas mandelii* [11], *Lactobacillus plantarum* [12], and *Bacillus halodurans* [13]. However, limited information is still available on the structure and function of these enzymes [7–9].

Halocynthiibacter arcticus is a rod-shaped Gram-negative bacteria from the Arctic region, which could be a valuable resource for biotechnological applications [14,15]. Specifically, structural information of enzymes from this bacterium is largely unknown. Here, we report the identification, purification, and preliminary crystallographic analysis of a novel cold-active esterase (*Ha*Est1) from *H. arcticus*. The recombinant enzyme was purified, characterized, and crystallized for structural studies. Considering that cold-active esterases have great importance for industrial applications, *Ha*Est1 could be an interesting industrial enzyme with cold-active properties.

2. Materials and Methods

2.1. Chemicals and Columns

Nucleic acid modifying enzymes and DNA purification kits were purchased from New England BioLabs (Ipsdwich, MA, USA) and Intron Korea (Daejon, Korea). PD-10 column for dialysis and His-tag affinity column for purification were obtained from GE Korea (Seoul, Korea). All other reagents of high purity grade were obtained from Sigma-Aldrich (Goyang-si, Korea).

2.2. Gene Cloning and Protein Purification

Microbial culture of *Halocynthiibacter arcticus* (Korea Collection of Type Cultures, KCTC 42129) and purification of its chromosomal DNA were carried out as previously described [16]. The open reading frame of the *Ha*Est1 gene was amplified and cloned into a pET-21a vector (pET-*Ha*Est1) using the following primers with *NheI* and *XhoI* (forward primer: 5′-GTAACCGCTAGCATGACAGACCCACAG-3′, and reverse primer: 5′-GCTGACTCGAGTCAGAATTTCGCCCG-3′). *E. coli* BL21(λDE3) cells were transformed and grown in LB medium at 18 °C with 1 mM isopropyl-β-D-1-thiogalactoside (IPTG) induction. Then, bacterial cells were centrifuged and resuspended in a cell disruption buffer (20 mM Tris-HCl pH 8.5, 150 mM NaCl, 20 mM imidazole). Following cell lysis by ultrasonication, supernatants were loaded onto a His-Trap affinity column. *Ha*Est1 was eluted by a gradient of imidazole method (from 50 mM to 200 mM). The fractions were buffer-changed with PD-10 column. The final proteins were collected and stored at −20 °C.

2.3. Activity Measurement of HaEst1

Zymographic assay of *Ha*Est1 was performed as previously described [17–19]. Protein purities and concentrations of *Ha*Est1 were determined using SDS-PAGE and Bio-Rad Protein Assay kit. Enzymatic activity of *Ha*Est1 was studied using 4-methylumbelliferyl (4-MU) acetate and 4-methylumbelliferyl (4-MU) phosphate as substrates. The fluorescence of 4-methylumbelliferone was observed in an Eppendorf tube using UV incubation box. Intrinsic fluorescence was measured using a Jasco FP-6200 spectrofluorometer (Easton, MD, USA).

2.4. Crystallization Method

For effective crystallization trials, purified *Ha*Est1 (20.0 mg/mL) was used with crystallization kits including MCSG 1T~4T (Anatrace), JCSG-plus (Molecular Dimensions), and PGA Screen (Molecular Dimensions) [20,21]. The screening process was carried out using an automated crystallization robot (SPT Labtech, San Diego, CA, USA). Initial droplets contained 300 nL of protein solution mixed with 300 nL of reservoir solution by sitting-drop vapor-diffusion method in a 96-well plate. Various crystals of *Ha*Est1 appeared in several conditions within a week. Diffraction-quality single crystals of *Ha*Est1 were observed under JCSG-Plus #16 condition of 10% (*w*/*v*) PEG 8000/8% ethylene glycol, 0.1 M Hepes-NaOH, pH 7.5.

2.5. Data Collection and Processing

A single crystal of *Ha*Est1 was removed and transferred to a cryo-protectant solution. After gentle soaking, the crystal was mounted on a synchrotron facility at beamline 5C of the Pohang Light Source (PAL, Pohang, Korea). X-ray diffraction data of *Ha*Est1 were collected at 100 K using the Eiger X 9M detector (Dectris, Switzerland). For complete X-ray diffraction data collection, the cryo-cooled crystal was rotated throughout 360° rotation with 1° oscillation per frame. Finally, collected X-ray data were processed and indexed using HKL2000 (see Table 1). Sequences of *Ha*Est1 and other enzymes were obtained from a public NCBI server, and multiple sequence alignment was prepared using ESPript.

Table 1. Data collection and processing statistics of *Ha*Est1.

Data Collection	
Synchrotron source	PAL-Korea, beamline 5C
Wavelength (Å)	0.9794
Temperature (K)	100
Detector	Eiger X 9M
Crystal to detector distance (mm)	400
Rotation range per image (°)	1
Exposure time per image (s)	1
Rotation range (°)	360
Data Processing	
Crystal parameters	
Space group	$P2_1$
No. of molecules in asymmetric unit	2
Unit-cell parameters (Å, °)	A = 35.69, b = 91.21, c = 79.15, α = 90, β = 96.91, γ = 90
Data Statistics	
Resolution range (Å)	50.00–2.10 (2.14–2.10)
Total no. of collected reflections	186,387 (122)
No. of unique reflections	29,269 (1435)
Completeness (%)	99.0 (99.6)
Multiplicity	6.4 (0.1)
$\langle I/\sigma(I)\rangle$	47.8 (7.67)
R_{merge} (%) †	8.1 (36.3)
R_{meas} (%) ‡	8.8 (39.8)
CC(1/2) (%)	99.1 (94.7)
Overall B-factor from Wilson plot (Å^2)	39.7

† $R_{merge} = \sum_{hkl} |I - \langle I\rangle| / \sum_{hkl} I$. ‡ $R_{meas} = \sum_{hkl} \{N(hkl)/[N(hkl) - 1]\}^{1/2} \sum_i |I_i(hkl) - \langle I(hkl)\rangle| / \sum_{hkl}\sum_i I(hkl)$.

3. Results and Discussion

An open reading frame encoding a novel cold-active esterase (*Ha*Est1, locus tag: WP_039000957, 756 bp) was detected and obtained from the *H. arcticus* chromosome. *Ha*Est1 has 252 amino acids with a pI of 5.21. Multiple sequence alignments of *Ha*Est1 with three related proteins indicated that *Ha*Est1 showed significant sequence identities with a putative hydrolase from *Agrobacterium vitis* (3LLC, 40.0%), a new family of carboxyl esterase with an OsmC domain from *Rhodothermus marinus* (5CML, 19.8%) [22], and a cinnamoyl esterase from *Lactobacillus johnsonii* LJ0536 (3PF8, 20.6%) [23] (Figure 1). Interestingly, *Ha*Est1 showed substantial sequence similarity to an alpha/beta hydrolase domain-containing protein 10 (ABHD10) from *Mus musculus* (6NY9, 23.8%). Catalytic triad of Ser-His-Asp as well as G-x-S-x-G motif were also conserved in these two proteins. This ABHD10 was recently shown to be an *S*-depalmitoylase affecting reduction/oxidation homeostasis [24].

Highly conserved catalytic residues of Ser^{101}, Asp^{198}, and His^{228} were identified, with Ser^{101} located in a typical GXSXG motif. Sequence analysis revealed that *Ha*Est1 has a high number of small amino acids such as Gly (11.9%) and Ala (9.9%). In addition, high percentages of Leu (9.1%) and Thr (7.9%) were also observed. The percentage of acidic amino acids (Asp + Glu) was larger than that of basic amino acids (Arg + Lys).

The recombinant *Ha*Est1 was purified using an immobilized His-tag metal-binding column (Figure 2A). The enzymatic activity of *Ha*Est1 was examined using 4-methylumbelliferyl (4-MU) acetate. As shown in Figure 2B, strong fluorescence due to hydrolysis reaction was observed for 4-MU acetate, although little hydrolysis reaction was carried out for control or 4-MU phosphate.

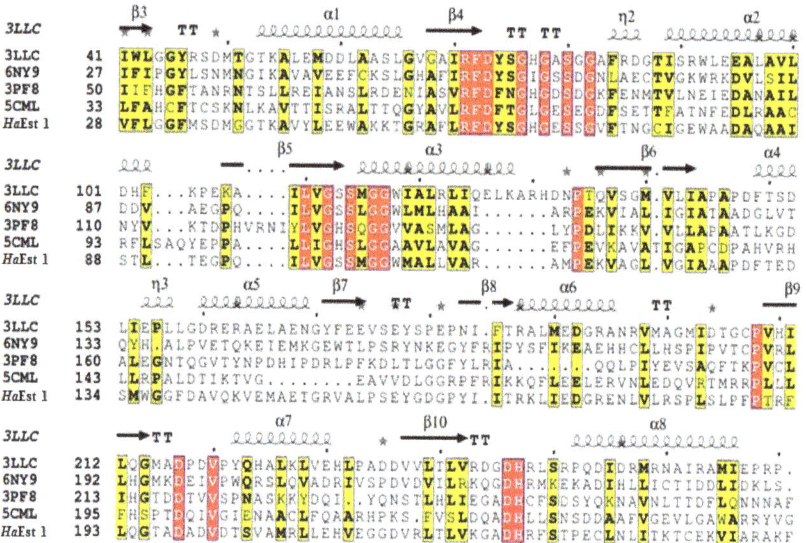

Figure 1. Multiple sequence alignments of *Ha*Est1. Identical and highly conserved amino acids among these proteins are displayed in red and yellow.

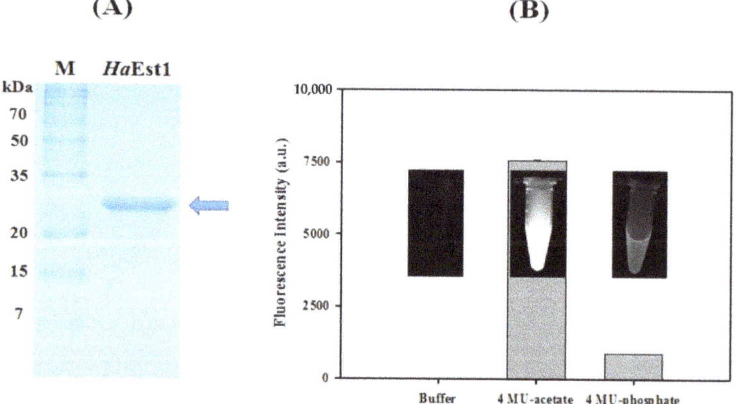

Figure 2. Purification and hydrolytic activity of *Ha*Est1. (**A**) SDS-PAGE analysis of *Ha*Est1. Arrow indicates the position of purified *Ha*Est1. (**B**) Hydrolysis of 4-MU acetate was examined with *Ha*Est1. Strong fluorescence was observed due to hydrolysis reaction by *Ha*Est1.

The diffraction-quality crystals grew to final dimensions of 0.6 × 0.5 × 0.2 mm within three days at 297 K (Figure 3), which were transferred to a paratone oil, cryo-protectant solution. The diffraction data set of *Ha*Est1 was indexed to $P2_1$ space group with unit cell parameter of a = 35.69 Å, b = 91.21 Å, c = 79.15 Å, and β = 96.9°. The final data were processed using HKL2000 to 2.10 Å resolution with 99.9% completeness. The final R_{merge} and $R_{p.i.m}$ values were 8.1% and 3.5% for 50–2.10 Å resolution. The final data collection statistics are summarized in Table 1. Assuming two molecules of *Ha*Est1 per asymmetric unit, Matthews coefficient (V_M) and solvent content were calculated to be 2.36 Å^3/Da and 47.9% [25].

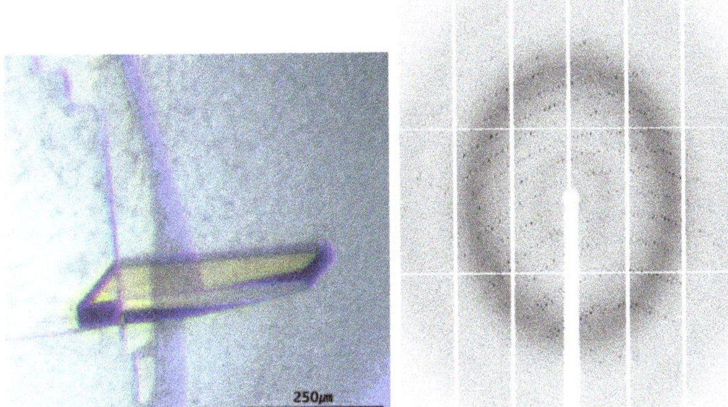

Figure 3. Representative crystal image (**left**) and diffraction pattern (**right**) of *Ha*Est1. The crystal has final dimensions of 0.6 × 0.5 × 0.2 mm.

We tried to solve the *Ha*Est1 structure using the molecular replacement method usingMOLREP [26]. The crystal structure of a putative hydrolase from *Agrobacterium vitis* (PDB code 3LLC) was used as a search model for the cross-rotation search. For the cross-rotation function calculation, we used data in the resolution range of 39.44–2.41 Å. The results of the cross-rotation showed that the highest peak height was above 7.49σ. The highest peak solution of the rotation function was used for the following translation function search. The solution model also gave a strong correlation coefficient value (above 0.52) in the translation function. Rigid body refinements and individual restrained B-factor refinements were performed using REFMAC5 [27]. After these refinement steps, an interpretable electron density map was calculated, as shown in Figure 4. Model building and further refinement are now underway. Furthermore, substrate or product bound *Ha*Est1 structure determination with site-directed mutagenesis experiments will be performed. Thus, a detailed structural analysis of *Ha*Est1 and protein engineering results will be published in an upcoming research paper.

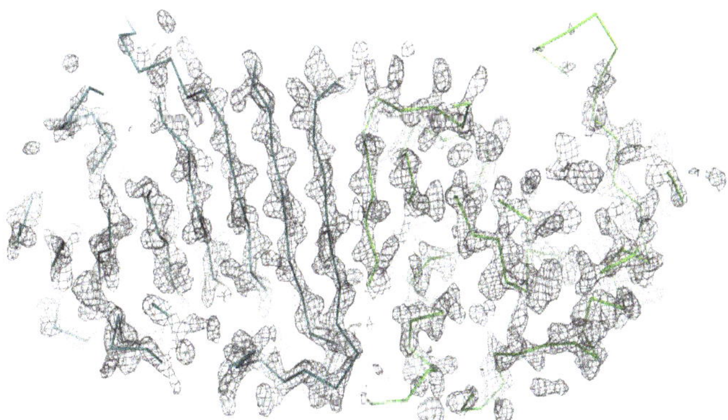

Figure 4. Electron density map of a portion of *Ha*Est1 structure contoured at 2σ. The initially built Cα trace is shown in green and deep teal color for each chain. The figure was made with Pymol.

Determination of the *Ha*Est1 structure will allow direct comparison to other mesophilic or thermophilic esterases, which could provide molecular insights on its reaction mechanism as well as its functional properties. In summary, this work deals with the purification and crystallographic analysis of a novel cold-active *Ha*Est1, which could be used for biotechnological applications. Furthermore, *Ha*Est1 could be further improved/mutated through protein or genetic engineering for its useful applications.

Author Contributions: S.J., W.Y. and J.W.C. identified and purified *Ha*Est1. J.H., H.D. and H.-W.K. got crystals and diffraction data. J.H.L., K.K.K. and T.D.K. wrote the manuscript. All authors have read and agreed to the published version of the manuscript.

Funding: This work was supported by Medical Research Center Program funded by the Korea government (MSIP) (No. 2011-0030074).

Data Availability Statement: Not applicable.

Conflicts of Interest: All authors declare no conflicts of interests.

References

1. Holmquist, M. Alpha/Beta-hydrolase fold enzymes: Structures, functions and mechanisms. *Curr. Protein Pept. Sci.* **2000**, *1*, 209–235. [CrossRef] [PubMed]
2. Gupta, R.; Gupta, N.; Rathi, P. Bacterial lipases: An overview of production, purification and biochemical properties. *Appl. Microbiol. Biotechnol.* **2004**, *64*, 763–781. [CrossRef] [PubMed]
3. Bhatt, P.; Bhatt, K.; Huang, Y.; Lin, Z.; Chen, S. Esterase is a powerful tool for the biodegradation of pyrethroid insecticides. *Chemosphere* **2020**, *244*, 125507. [CrossRef]
4. Sarmah, N.; Revathi, D.; Sheelu, G.; Yamuna-Rani, K.Y.; Sridhar, S.; Mehtab, V.; Sumana, C. Recent advances on sources and industrial applications of lipases. *Biotechnol. Prog.* **2018**, *34*, 5–28. [CrossRef]
5. Chen, Y.; Black, D.S.; Reilly, P.J. Carboxylic ester hydrolases: Classification and database derived from their primary, secondary, and tertiary structures. *Protein Sci.* **2016**, *25*, 1942–1953. [CrossRef]
6. Oh, C.; Kim, T.D.; Kim, K.K. Carboxylic ester hydrolases in bacteria: Active site, structure, function and application. *Crystals* **2019**, *9*, 597. [CrossRef]
7. Santiago, M.; Ramírez-Sarmiento, C.A.; Zamora, R.A.; Parra, L.P. Discovery, molecular mechanisms, and industrial applications of cold-active Enzymes. *Front. Microbiol.* **2016**, *7*, 1408. [CrossRef]
8. Siddiqui, K.S. Some like it hot, some like it cold: Temperature dependent biotechnological applications and improvements in extremophilic enzymes. *Biotechnol. Adv.* **2015**, *33*, 1912–1922. [CrossRef]
9. Dumorné, K.; Córdova, D.C.; Astorga-Eló, M.; Renganathan, P. Extremozymes: A potential source for industrial applications. *J. Microbiol. Biotechnol.* **2017**, *27*, 649–659. [CrossRef]
10. Park, S.-H.; Yoo, W.; Lee, C.W.; Jeong, C.S.; Shin, S.C.; Kim, H.-W.; Park, H.; Kim, K.K.; Kim, T.D.; Lee, J.H. Crystal structure and functional characterization of a cold-active acetyl xylan esterase (PbAcE) from psychrophilic soil microbe *Paenibacillus* sp. *PLoS ONE* **2018**, *13*, e0206260. [CrossRef] [PubMed]
11. Truongvan, N.; Jang, S.H.; Lee, C. Flexibility and stability trade-off in active site of cold-adapted *Pseudomonas mandelii* esterase EstK. *Biochemistry* **2016**, *55*, 3542–3549. [CrossRef] [PubMed]
12. Esteban-Torres, M.; Mancheño, J.M.; Rivas, B.D.L.; Muñoz, R. Characterization of a cold-active esterase from *Lactobacillus plantarum* suitable for food fermentations. *J. Agric. Food Chem.* **2014**, *62*, 5126–5132. [CrossRef]
13. Noby, N.; Hussein, A.; Saeed, H.; Embaby, A.M. Recombinant cold -adapted halotolerant, organic solvent-stable esterase (estHIJ) from *Bacillus halodurans*. *Anal. Biochem.* **2020**, *591*, 113554. [CrossRef]
14. Lee, Y.M.; Baek, K.; Lee, J.; Lee, H.K.; Park, H.; Shin, S.C. Complete genome sequence of Halocynthiibacter arcticus PAMC 20958(T) from an Arctic marine sediment sample. *J. Biotechnol.* **2016**, *224*, 12–13. [CrossRef]
15. Baek, K.; Lee, Y.M.; Shin, S.C.; Hwang, K.; Hwang, C.Y.; Hong, S.G.; Lee, H.K. *Halocynthiibacter arcticus* sp. nov., isolated from Arctic marine sediment. *Int. J. Syst. Evol. Microbiol.* **2015**, *65*, 3861–3865. [CrossRef] [PubMed]
16. Jeon, S.; Hwang, J.; Yoo, W.; Do, H.; Kim, H.-W.; Kim, K.K.; Lee, J.H.; Kim, T.D. Crystallization and preliminary x-ray diffraction study of a novel bacterial homologue of mammalian hormone-sensitive lipase (*Ha*Lip1) from *Halocynthiibacter arcticus*. *Crystals* **2020**, *10*, 963. [CrossRef]
17. Le, L.T.H.L.; Yoo, W.; Lee, C.; Wang, Y.; Jeon, S.; Kim, K.K.; Lee, J.H.; Kim, T.D. Molecular characterization of a novel cold-active hormone-sensitive lipase (*Ha*HSL) from *Halocynthiibacter arcticus*. *Biomolecules* **2019**, *9*, 704. [CrossRef] [PubMed]
18. Jang, E.; Ryu, B.H.; Shim, H.W.; Ju, H.; Kim, D.W.; Kim, T.D. Adsorption of microbial esterases on Bacillus subtilis-templated cobalt oxide nanoparticles. *Int. J. Biol. Macromol.* **2014**, *65*, 188–192. [CrossRef] [PubMed]
19. Wang, Y.; Le, L.T.H.L.; Yoo, W.; Lee, C.W.; Kim, K.K.; Lee, J.H.; Kim, T.D. Characterization, immobilization, and mutagenesis of a novel cold-active acetylesterase (EaAcE) from *Exiguobacterium antarcticum* B7. *Int. J. Biol. Macromol.* **2019**, *136*, 1042–1051. [CrossRef]
20. Lee, C.W.; Yoo, W.; Park, S.H.; Le, L.T.H.L.; Jeong, C.S.; Ryu, B.H.; Shin, S.C.; Kim, H.W.; Park, H.; Kim, K.K.; et al. Structural and functional characterization of a novel cold-active S-formylglutathione hydrolase (SfSFGH) homolog from *Shewanella frigidimarina*, a psychrophilic bacterium. *Microb. Cell Factories* **2019**, *18*, 140. [CrossRef]

21. Lee, C.W.; Lee, S.; Jeong, C.S.; Hwang, J.; Chang, J.H.; Choi, I.G.; Kim, T.D.; Park, H.; Kim, H.Y.; Lee, J.H. Structural insights into the psychrophilic germinal protease PaGPR and its autoinhibitory loop. *J. Microbiol.* **2020**, *58*, 772–779. [CrossRef] [PubMed]
22. Jensen, M.V.; Horsfall, L.E.; Wardrope, C.; Togneri, P.D.; Marles-Wright, J.; Rosser, S.J. Characterisation of a new family of carboxyl esterases with an OsmC domain. *PLoS ONE* **2016**, *11*, e0166128. [CrossRef]
23. Lai, K.K.; Stogios, P.J.; Vu, C.; Xu, X.; Cui, H.; Molloy, S.; Savchenko, A.; Yakunin, A.; Gonzalez, C.F. An inserted α/β subdomain shapes the catalytic pocket of *Lactobacillus johnsonii* cinnamoyl esterase. *PLoS ONE* **2011**, *6*, e23269. [CrossRef]
24. Cao, Y.; Qiu, T.; Kathayat, R.S.; Azizi, S.A.; Thorne, A.K.; Ahn, D.; Fukata, Y.; Fukata, M.; Rice, P.A.; Dickinson, B.C. ABHD10 is an S-depalmitoylase affecting redox homeostasis through peroxiredoxin-5. *Nat. Chem. Biol.* **2019**, *15*, 1232–1240. [CrossRef] [PubMed]
25. Matthews, B.W. Solvent content of protein crystals. *J. Mol. Biol.* **1968**, *33*, 491–497. [CrossRef]
26. Vagin, A.; Teplyakov, A. Molecular replacement with MOLREP. *Acta Crystallogr. Sect. D Biol. Crystallogr.* **2010**, *66*, 22–25. [CrossRef] [PubMed]
27. Murshudov, G.N.; Skubák, P.; Lebedev, A.A.; Pannu, N.S.; Steiner, R.A.; Nicholls, R.A.; Winn, M.D.; Long, F.; Vagin, A.A. REFMAC5 for the refinement of macromolecular crystal structures. *Acta Crystallogr. Sect. D Biol. Crystallogr.* **2011**, *67*, 355–367. [CrossRef] [PubMed]

Article

Molecular Dynamics Investigation of Phenolic Oxidative Coupling Protein Hyp-1 Derived from *Hypericum perforatum*

Joanna Smietanska *, Tomasz Kozik, Radoslaw Strzalka, Ireneusz Buganski and Janusz Wolny

Faculty of Physics and Applied Computer Science, AGH University of Science and Technology, 30-059 Krakow, Poland; tomasz.kozik@fis.agh.edu.pl (T.K.); strzalka@fis.agh.edu.pl (R.S.); buganski@fis.agh.edu.pl (I.B.); wolny@fis.agh.edu.pl (J.W.)

* Correspondence: joanna.smietanska@fis.agh.edu.pl

Abstract: Molecular dynamics (MD) simulations provide a physics-based approach to understanding protein structure and dynamics. Here, we used this intriguing tool to validate the experimental structural model of Hyp-1, a pathogenesis-related class 10 (PR-10) protein from the medicinal herb *Hypericum perforatum*, with potential application in various pharmaceutical therapies. A nanosecond MD simulation using the all-atom optimized potentials for liquid simulations (OPLS–AA) force field was performed to reveal that experimental atomic displacement parameters (ADPs) underestimate their values calculated from the simulation. The average structure factors obtained from the simulation confirmed to some extent the relatively high compliance of experimental and simulated Hyp-1 models. We found, however, many outliers between the experimental and simulated side-chain conformations within the Hyp-1 model, which prompted us to propose more reasonable energetically preferred rotameric forms. Therefore, we confirmed that MD simulation may be applicable for the verification of refined, experimental models and the explanation of their structural intricacies.

Keywords: molecular dynamics simulation; Hyp-1 protein; rotamers; B-factors; thermal motions

Citation: Smietanska, J.; Kozik, T.; Strzalka, R.; Buganski, I.; Wolny, J. Molecular Dynamics Investigation of Phenolic Oxidative Coupling Protein Hyp-1 Derived from *Hypericum perforatum*. *Crystals* **2021**, *11*, 43. https://doi.org/10.3390/cryst11010043

Received: 22 November 2020
Accepted: 1 January 2021
Published: 6 January 2021

Publisher's Note: MDPI stays neutral with regard to jurisdictional claims in published maps and institutional affiliations.

1. Introduction

Since the first successful macromolecular simulation using molecular dynamics (MD) methods was performed in 1977 [1], MD simulations have grown into one of the most powerful tools for understanding the physical basis of the structure of proteins [2], their biological functions and the role of different types of interactions within macromolecules [3–6]. Over the past 40 years, timescales that can be covered by atomistic MD simulation are growing faster than Moore's law [7]. Supercomputers have been built for studies on protein folding via large-scale simulation [8]. In response to the growing interest in physics-based methods of protein simulations during the current decade, the most popular MD simulation packages such as CHARMM [9], NAMD [10], GROMACS [11] and AMBER [12] have improved their atomistic simulation algorithms, computing performance, as well as their methods of comprehensive analysis and experimental validation of underlying physical models [13–15]. The calculation of the conformational energy landscape, approachable to protein molecules under certain force fields, connects information about structure and dynamics arising from the internal motion of molecules [16]. A meticulous study of individual atomic motions as a function of time provides insight into biomolecular properties of modeled systems, especially those which are difficult to verify experimentally by methods of X-ray crystallography or NMR spectroscopy [17]. As an excellent tool for studying specific interactions ruling the behavior of biological macromolecules, MD simulations have proven their effectiveness in the fields of protein folding [18–20], conformational change analysis [21], ligand binding [22] and ab initio structure refinement [23]. The proper interpretation of the results of MD simulation requires the consideration of some practical issues and artifacts. Apart from accidental programming and user errors or algorithmic issues, various known problems include the choice of force field, the impact of thermostatting, the "flying ice cube" effect, non-physical behavior of water molecules, method of electrostatic interactions computing or treatment of Lorentz–Berthelot rules for simulations of a mixture of different atoms [24]. Insufficient sampling can also limit applications of MD simulations due to the generation of rough energy landscapes with many local minima separated by high-energy barriers, making it easy to fall into a non-functional state that is hard to jump out of in most conventional simulations [25].

To date, nearly 85% of protein structures deposited in the RCSB Protein Data Bank (PDB) were solved using X-ray crystallographic methods. However, this technique provides only static, time- and ensemble-average representation of the molecular arrangement in a crystal [26], ignoring the fact that atomic motions never disappear completely, even at low temperatures [27]. Typically, individual isotropic/anisotropic atomic displacement parameters (ADPs, *B*-factors or temperature factors) described as

$$B = 8\pi^2 < u^2 >, \quad (1)$$

where $< u^2 >$ is the mean square atomic displacement of a given atom from the ideal position (called root-mean-square fluctuations, RMSF), reflect atoms' oscillation amplitude around their equilibrium positions in a crystal structure. As experimental ADPs are strongly correlated with atomic form factors, electron density spreading during atomic motions results in a strong decrease in atom scattering power. The analysis of ADPs provides a range of information about the paths of thermal-controlled motions within protein cavities, model quality, preferred side-chain conformations, protein thermal stability, and local flexibility [28,29].

Potential static or dynamic conformational disorder, modeling, and parameterization errors may influence ADPs which makes their distribution a sensitive indicator of the degree of crystal disorder and model accuracy [30]. However, it may be difficult to distinguish actual thermal motions from positional dispersion affected by lattice disorder. Differences in ADP distribution along with protein crystal structure can imply regions of high thermal mobility, such as less stable side-chain conformations or crystal moieties [17,31]. While ADPs are not experimental observables and we do not have our high-resolution diffraction data, restraints assuming the isotropy or anisotropy of ADPs must be imposed on the model [32]. This simplification may lead to a significant deviation of experimental ADPs from reality and results in a growing number of PDB-deposited structures with unreliably high B factors. A possible way of considering correlated atomic motions in protein comes from using the translation–libration–screw model (TLS) that treats proteins as rigid bodies [33,34]. The key problem is the selection of the number of TLS groups and their range, which is typically performed based on the chemical knowledge of the rigidity of certain groups of atoms [35].

To verify existing protein crystal structures and avoid the unrealistic modeling of molecules, the urgent need to develop a model predicting reliable ADPs' values has been growing [27,36]. Recently, it has been proven that the global distribution of ADPs is described by shifted inverse-gamma distribution (SIGD). The SIGD model demands a definition of three essential parameters: shape (α), scale (β), and shift (B_0) of the distribution [28]. Starting values are iteratively improved in Fisher's scoring method to obtain maximum convergence as well as the physical reliability of the estimated distribution parameters. The validation of macromolecular models proceeds through the juxtaposition of ADPs distribution from an existing model with a contour plot based on the calculated SIGD parameters. If ADP distribution for the whole structure or individual domains/chains obey particularly different SIGDs, such multimodality can indicate the presence of incorrectly modeled parts of molecules that require rigorous inspection.

The structure of proteins is mainly determined by different types of bonding interactions between the side-chain groups of the amino acids [37]. Some specific combinations of dihedral angles corresponding to given conformations of side-chain rotational isomers, so-called rotamers, are preferred [38,39]. In many experimental structures, part of the side-chain conformations is unfavorable, as indicated by the improper stereochemistry or unusually high ADP values [40]. When the definition of a side chain in electron density maps is blurred or local ADP parameters are suspiciously large, protein dynamics simulations may be helpful in the verification of the energetically favored local conformations or identification of non-rotameric states [41].

As the subject of our research, the model of the structure of the Hyp-1 protein obtained from the *Hypericum perforatum* medicinal herb commonly known as St. John's wort was considered. The healing properties of St. John's wort preparations have been known for millennia, with their main active component hypericin, a red-colored quinone derivative, acting as a remedy for depression [42]. As a light-sensitive compound, hypericin is also used in photodynamic anticancer and antiviral therapies [43]. It was initially suspected that hypericin synthesis from another natural anthraquinone, emodin, occurs in a complicated multi-step dimerization reaction catalyzed by the Hyp-1 protein. The analysis of sequence similarities (~50%) allows for the classification of Hyp-1 as a plant pathogenesis-related class 10 (PR-10) protein [44]. The PR-10 family proteins are typically produced in plants as a response to stressful environmental factors such as drought, salinity, or pathogen invasion. The exact biological activity of most of the PR-10 proteins remains unknown and widely disputed, which stands in opposition to their well described characteristic folding canon [43,44]. The presence of this hydrophobic pocket strongly suggests the ability of the Hyp-1 protein to bind various biological ligands such as melatonin [45]. Recent studies of Hyp-1 in the complex with fluorescent dye 8-anilinonaphthalene-1-sulfonate (ANS) revealed the formation of unique commensurately modulated and tetartohedrally twinned macromolecular superstructures [46]. In the present structure, the protein cavity is occupied by polyethylene glycol molecules of different chain lengths from the polyethylene glycol (PEG400) cryoprotectant solution. Although Hyp-1, as many other PR-10 proteins, remains monomeric in solution, crystallization under oxidating conditions in the complex with PEG leads to the formation of a compact dimer formed via covalent S–S linking involving Cys126 and the parallel association of the β1 strands with another Hyp-1 partner molecule [43].

A high-resolution Hyp-1 protein structure (PDB entry 3IE5) was solved at the P2_12_12_1 space group. Despite the high stereochemical quality of the model confirmed by the main-chain torsion distribution on the Ramachandran plot, the validation report still indicates that 1.8% of the side-chain conformations are classified as outliers, which is a common problem for protein structures. The overall clash score was equal to 10 which means that even 1% of atoms are involved in too-close contacts with neighboring residues.

We distinguished a physics-based approach such as MD simulation monitoring direct, atomic motions in the unit cell and knowledge-based X-Ray crystallography imposing many restraints as a complementary help to optimize a model even if a refinement process seems to be finished with some unresolved conformational problems. We also want to answer a question: to what extent our arbitrary knowledge of ADPs distribution strongly depends on restraints applied during refinement can be checked by the simulation of direct, individual atomic motions in the unit cell?

In the presented paper, we used pico- and nanoscale MD simulations as an energetically based auxiliary tool for the verification and improvement of the experimental protein structure model. We chose a high-resolution crystal structure model of Hyp-1 protein from St. John's wort with potential in various medicinal therapies. The comparison between experimental ADPs and those calculated directly from atomic motions during the simulation was performed to check the validity of thermal factors distribution from the Hyp-1 structure. The side-chain angle distributions for different types of amino acids (non-polar, polar, and aromatic) were measured to find energetically preferred rotamer forms, especially for conformations poorly visible in electron density maps or partially occupied. The verification of the average simulated protein structure was conducted by comparison between experimental, refined, and calculated from the MD model structure factors.

2. Materials and Methods

2.1. *Hyp-1 Structure Model*

As starting coordinates, we used PDB entry 3IE5 from *Hypericum perforatum* plant protein Hyp-1 at 1.69 Å resolution refined with R_{work}/R_{free} equal to 17.0/20.6%, respectively. The asymmetric part of the unit cell contains two independent protein molecules (A and B) forming a dimer through hydrogen bond interactions between their β-sheets. Chain A consists of 160 amino acids (159 residues labeled Met1-Ala159 from the main chain and artificial Thr-1 molecule as cloning artifact), whereas molecule B has 164 amino acids including 5 additional expression tags (Ile-5 ... Thr-1). In addition to proteins, the starting model contained different polyethylene glycol molecules from the PEG400 cryoprotectant solution bound within the internal cavity and on the protein surface. For MD simulation purposes, the whole structure model was used, but during the data analysis, the uninteresting glycol residues were omitted due to their high flexibility and relatively poor embedding in electron density maps. The final asymmetric unit cell used for the MD simulation contains two molecules of Hyp-1, 258 water molecules, a single Cl^- anion located by Lys56A residue, and 10 polyethylene glycol molecules of different chain lengths. The Hyp-1 molecules A and B bind in the internal cavities and 2 and 3 PEG molecules, respectively. Further PEG moieties are located at the protein surface mostly by interactions between protonated lysine and PEG molecules.

2.2. *Molecular Dynamics Simulation*

All molecular dynamics simulations were performed in the free and open source simulation package GROMACS 2019.3 [47–52]. The general purpose OPLS–AA force field was used [53–61] in all simulations. The TIP4P-Ew water model [62] was used. All Glu and Asp residues were treated as negatively charged and all Lys and Arg residues as positively charged, whereas the other residues were assumed to be neutral. The internal residue database of the simulation package was complemented with the topologies of the various glycols present in the simulated system. All of the necessary potentials were already defined in the force field. Partial charges of the atoms within glycol molecules were obtained by fitting point charges to the electrostatic potential calculated from the charge density obtained by DFT [63,64] calculations using the 6-311G** Gaussian type orbital basis set [65] with the B3LYP exchange-correlation potential [66,67] in the NWChem package [68].

The starting model for the series of simulations was the structure of Hyp-1 dimer as published in [43] with four units in the asymmetric unit duplicated in each direction (x, y, z) to form a $2 \times 2 \times 2$ supercell of the structure. Therefore, the final system included 64 individual protein copies and 208,032 atoms in a 67.45 Å by 137.85 Å by 215.27 Å rectangular box.

The starting structure underwent an energy minimization (EM) procedure that quickly converged without significantly altering the structure (Table 1). Then, an equilibrating NVT simulation was performed. It consisted of 100,000 steps of 1 fs each, in which the temperature was coupled to 292 K (independently for groups of protein and non-protein residues) using a velocity rescaling thermostat with a coupling constant of 0.1 ps, short-range interactions (van der Waals and Coulomb) were cut off at a distance of 3.2 nm, long-range interactions were handled using the PME method [69,70], the Verlet cut off scheme [71] was used and the neighbor list was updated every 20 steps. Periodic boundary conditions were used in all three dimensions and the initial velocities of atoms were generated from a Maxwell distribution at 292 K.

Table 1. Simulation steps with their IDs and the number of steps/time step.

Simulation Step	ID	Number of Steps/Time Step (fs)
Energy minimization	EM	100,000/1
NVT equilibration	NVT	100,000/1
NPT equilibration	NPT	1,000,000/1
Simulation of rotamer 1	ROT1	100/200
Simulation of rotamer 2	ROT2	10/20
Simulation of rotamer 3	ROT3	1/2
Cooling down	CD	1,200,000/1
RMSF gathering	RMSF	1,000,000/1

The output structure from the first equilibrating simulation was further equilibrated in an NPT simulation of 1,000,000 steps of 1 fs each. The settings were the same as in the earlier simulation. In addition to those, an isotropic Parrinello–Rahman barostat with a coupling constant of 2 ps was applied to couple the pressure to 1 bar, with compressibility set to 4.5×10^{-5} bar^{-1}.

The equilibrated structure obtained in the above simulation sequence was the basis of performing additional simulations with identical settings used for gathering data regarding atomic positions in 100 ps, 10 ps, and 1 ps time scales, consisting of a corresponding number of 1 fs time steps. The positions of atoms were written to the trajectory file every 200 fs, 20 fs, and 2 fs accordingly. These data were used for calculating the angles corresponding to rotamers (ROTs).

The equilibrated structure was cooled down to 100 K (temperature of X-Ray crystallographic measurements) in yet another simulation of 1,200,000 steps of 1 fs each (CD). The settings were the same as earlier, except for the coupling temperature, which varied in the course of the simulation. The temperature stayed at 292 K for the first 100 ps of the simulated time, was linearly decreased to 100 K during the next 1 ns of simulated time and stayed at 100 K for the final 100 ps of the simulated time. Although cooling too quickly can trap side-chain conformations in unnatural states, it can be prevented by a decelerated linear decrease in temperature during 1 ns as adopted in our CD protocol, which provides sufficient time for adaptive changes in the protein.

Finally, the cooled down structure underwent a simulation of 1,000,000 steps 1 fs each (RMSF) for gathering atom-positional RMSF data. Positions of atoms were written to the trajectory file every 2 ps.

2.3. Comparison between Experimental and Simulated Structure Model

We compared the experimental model marked as PDB entry 3IE5 and the simulated Hyp-1 structure to determine the extent ot which our simulation preserved the characteristic PR-10 fold of the protein. We also wanted to prevent some drastic conformational changes within the Hyp-1 model. For our average conformation calculated by superposing of 32 Hyp-1 dimers from the simulation box, structure factors were calculated using the phenix.fmodel from the PHENIX package [72]. Experimental (F_{exp}) and obtained from model structure factors (F_{calc}) were recovered from the MTZ file accompanying the deposited 3IE5 structure. To determine the relative error, we compared the simulated (F_{md}) structure factors with those calculated from the model and subsequently excluded those with an error percentage above 20%. Root-mean-square deviation (RMSD) of Cα atoms of experimental and simulated Hyp-1 model were calculated in *gesamt* [73] to highlight the similarity between both structures. Calculation of RMSD of Cα atoms was performed using the gesamt from the CCP4 package.

2.4. Application of MD Nanoscale Simulation to Calculate ADPs Distribution

ADPs of Cα and Cγ atoms from the MD simulation were calculated by averaging over all time steps for each of the 32 dimers from the $2 \times 2 \times 2$ simulation box under previously described simulation conditions. Atoms lying on the surface of the simulation box were omitted in the calculation of individual ADPs. By that, isotropic ADPs were calculated with respect to the average MD structure after equilibration using Equation (1) for bulk atoms in the simulation box. The next step was aligning all selected conformations with respect to the first saved one to find the average protein structure in the simulation box. The coordinates of the average conformation were found using the root mean square fit of all CA and CG or CG2 atoms. Therefore, calculated ADPs averaged over all conformations were the mean of all atomic ADPs from 32 protein dimers inside the simulation box. The convergence of ADPs from the partial data set including conformational changes during the simulation was assessed by RMSD calculation with its standard error (SE) according to:

$$SE = \sqrt{\sum (B_i - \overline{B})^2 / N(N-1)} \quad (2)$$

where B_i refers to an ADP of an i-th conformation, $\overline{B}$ is the ADP of the calculated average structure, and N is the number of conformations in the simulation box [74]. To determine MD-based ADPs not just from Equation (1), the average Hyp-1 model after the MD simulation was refined in 50 cycles with isotropic ADPs and using the TLS model with 5 TLS groups per each chain in Refmac [75].

2.5. Stereochemical Constraints of Main- and Side-Chain Conformations

Physics-based MD results were compared with the observed side-chain amino acid distribution in the experimental structure. An experimental PDB-derived model was refined using stereochemical restraints according to a standard library [76]. We excluded the risk of steric clashes restricting rotameric form using short-contact restraints with a cut-off limit at 3.2 nm. Partial occupancy of some side chains (Ile6A and B, Arg27A and B, Leu31A, Phe72A, Leu105B, Lys139B) was omitted during MD calculations with their higher occupied form approved. Additionally, we rejected residues introduced at the cloning stage. Allowed rotameric forms were determined based on The Richardson's (Son of Penultimate) Rotamer Library [77] used commonly as a dataset in the post-refinement verification of structure model stereochemistry, e.g., in Coot [78] and Refmac programs and implemented during the refinement of the experimental Hyp-1 model. Calculations of energetically preferred rotameric forms involve different groups of amino acids: nonpolar (Leu, Met, Pro, and Val), aromatic (Phe, Tyr, His), polar (Asn, Gln, Cys, Ser, and Thr), basic (Lys, Arg) and acidic (Asp, Glu) residues in both α-helix and β-sheet backbone conformations. Table 2 contains definitions of χ_i backbone dihedral angles used during the simulation to monitor backbone angle variations. Values of dihedral angles for all 16 residues considered in this paper cover the range from 0° to 360°, except for χ_2 angle in the case of Phe, Tyr, and His, which is connected with the possibility of aromatic ring inversion in these compounds [79].

Table 2. Definitions of the torsion angles used for the calculation of χ_i dihedral angles and the determination of side-chain conformation.

	Initial Torsion Angle	Intermediate Torsion Angles	Terminal Torsion Angle
Arg	N-CA-CB-CG	CA-CB-CG-CD CB-CG-CD-NE	CG-CD-NE-CZ
Asn	N-CA-CB-CG		CA-CB-CG-OD1
Asp	N-CA-CB-CG		CA-CB-CG-OD1
Cys	N-CA-CB-SG		
Gln	N-CA-CB-CG	CA-CB-CG-CD	CB-CG-CD-OE1
Glu	N-CA-CB-CG	CA-CB-CG-CD	CB-CG-CD-OE1
His	N-CA-CB-CG		CA-CB-CG-ND1
Leu	N-CA-CB-CG		CA-CB-CG-CD1
Lys	N-CA-CB-CG	CA-CB-CG-CD CB-CG-CD-CE	CG-CD-CE-NZ
Met	N-CA-CB-CG	CA-CB-CG-SD	CB-CG-SD-CE
Phe	N-CA-CB-CG	CA-CB-CG-CD1	
Pro	N-CA-CB-CG		
Ser	N-CA-CB-OG		
Thr	N-CA-CB-OG1		
Tyr	N-CA-CB-CG	CA-CB-CG-CD1	
Val	N-CA-CB-CG1		

2.6. Comparison between Experimental and Calculated Distributions of Side-Chain Dihedral Angles

The side-chain dihedral angles for each residue in the experimental Hyp-1 model were calculated using *rotamer* [77,80] from the CCP4i software package. By the analysis of the (Son of Penultimate) Rotamer database, the respective rotameric form has been assigned to each of them, and additionally, side-chain outliers were identified. The reference calculated dataset was created from additional simulations following structure equilibration performed to monitor the positions of atoms at each 100, 10, and 1 ps timescales with the corresponding 200, 20, and 2 fs steps. For each residue, the χ_i torsion angles between respective atoms as mentioned in Table 1 were calculated. Obtained values for both α-helix and β-sheet backbone regions were compared with The Richardson's (Son of Penultimate) Rotamer Library content to identify predicted rotameric form. Model-based results were confronted with energetically favorable side-chain conformers from MD simulation. When some differences between the preferred rotameric form in the experimental and simulated structure were observed, we calculated the probability distribution of each side-chain dihedral angle $P(\chi_i)$, the number of samples in a small increment $\Delta\chi = 5°$ and then summed over all 60 protein conformations within a simulation box. Hence, we included the averaging of bond lengths, bond angles, and ω dihedral angles in different protein conformations. Each $P(\chi_i)$ distribution was normalized to fulfill the condition that $\int P(\chi_i)dx_i = 1$. To organize our analysis, we separately calculated the monomodal dihedral angles distribution $P(\chi_1)$ for Pro, Ser, Cys, Val and Thr residues, bimodal $P(\chi_1, \chi_2)$ distributions for the groups of Leu, Asp, Asn and the aromatic residues (Phe, His, Tyr), three- $P(\chi_1, \chi_2, \chi_3)$ or four-modal $P(\chi_1, \chi_2, \chi_3, \chi_4)$ distributions for residues with longer side chains, e.g., Glu, Gln, Met, Arg, and Lys.

3. Results and Discussion

3.1. Impact of MD Simulation Parameters and Possible Artifacts

In many previous studies of proteins, the cutoff of atom and residue contacts was selected arbitrarily in the range of 3.8–9.0 Å based on the properties of a particular system and the optimization of data processing, although some attempts to rationalize this process were presented [81–83]. In our simulation, long-range cutoff was used to not strongly affect the structural properties of the folded state as well as to provide simulation convergence in a given time. As the simple truncation of the electrostatic interactions at too short cutoff can create artificial boundary problems and even neglect important long-range interactions, we decided to use the value of 3.2 nm to maximally broaden the range of included van der Waals forces regarding accessible computational power and the large size of the protein [84]. The inclusion of longer-range electrostatic interactions is mostly limited by computational costs [85] and did not have a negative impact on our simulation. Moreover, neglecting long-range interactions may cause unacceptable large atomic fluctuations of protein residues, which cannot be explained in plain terms of environmental differences between X-ray studies and simulation. The simulation time of 1.2 ns for the longest trajectory was found to be sufficient to catch conformational changes and thermal motions in the Hyp-1 protein system. The convergence of simulation was confirmed by the observation of the atom-positional RMSD calculated against the experimental structure for each protein molecule over time and averaged over 64 individual protein copies inside the simulation box (Figure 1). Starting from experimental data, our system was not far from equilibrium, which results in the successful stabilization of RMSD values after 100 ps of the second 1 ns NPT simulation. Longer simulations are typically efficient in the much more complicated atomistic modeling of protein folding, structure dynamics, or protein-ligand interactions.

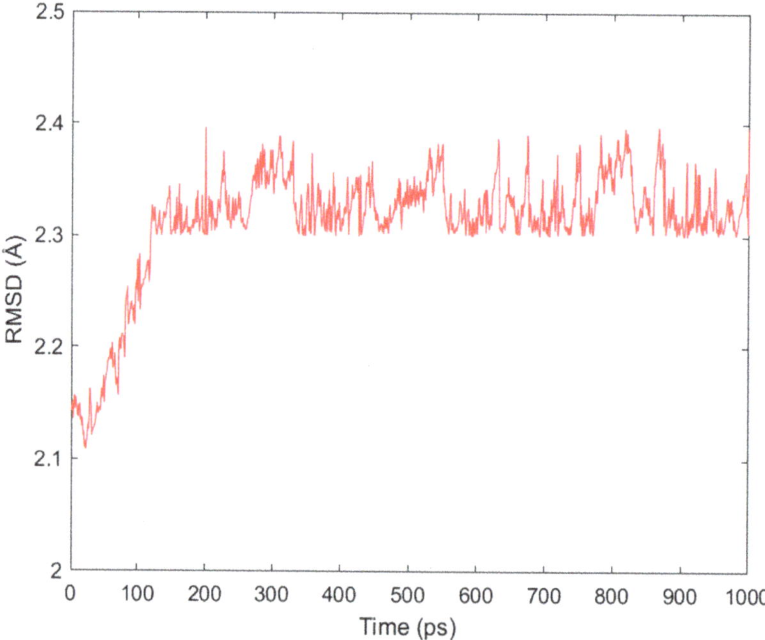

Figure 1. Average all-atom root-mean-square deviation (RMSD) calculated over time for the Hyp-1 protein structure during 1 ns of NPT equilibration. The values shown were averaged over 64 monomers in the crystal.

3.2. Accuracy of Experimental and Simulated Hyp-1 Structure Model

For the experimental Hyp-1 model, 39,745 unique reflections were collected, with 1590 of them used as a test set. Accordingly, we obtained a set of 39,737 reflections from the simulated average model using *phenix.fmodel*. A maximum relative error equal to 20% was applied to exclude those reflections for which structure factors exhibit high differences and low level of comparability. Therefore, 24,080 reflections have had an acceptable degree of compliance to be presented in the logarithmic scale on the respective F_{calc}/F_{md} and F_{md}/F_{exp} plots (Figure 2).

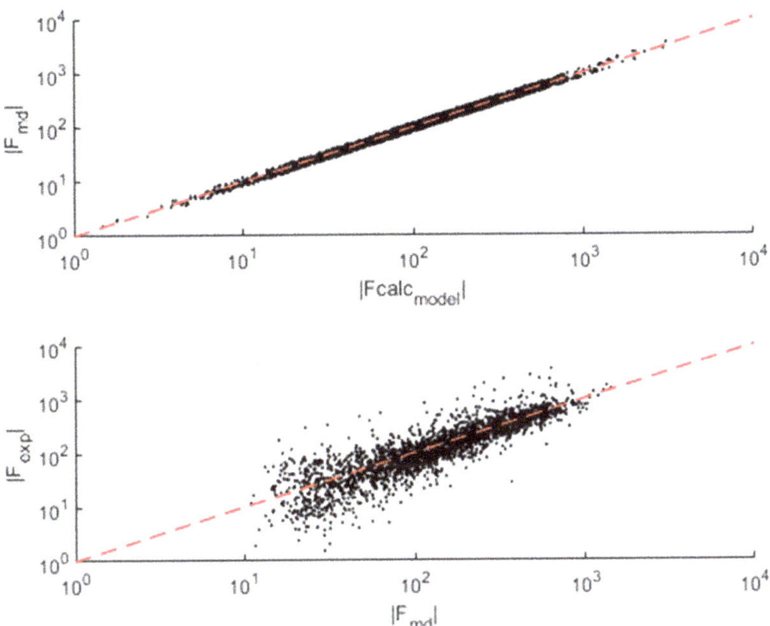

Figure 2. F_{calc}/F_{md} and F_{md}/F_{exp} plots for the Hyp-1 protein structure. The maximal permissible relative error between reflections with identical indices was established at a level of 20%. The red dashed line at 45° should serve as a guide to the eye and tool to visualize trends in changes of reflection intensity.

The calculated R-factor between all F_{md} and F_{calc} was equal to 10.6%, while the analogous one between F_{exp} and F_{md} data reaches 23.6%. Although reflections between experimental and simulated models have been fitted with only a small spread, we observed a blur of reflections on F_{md}/F_{exp} plot. This is apparently connected with an imperfect refinement of the structure model, or the multiple scattering phenomena, which was observed in many complex metallic alloy systems, e.g., quasicrystals [85–87]. Within the original Hyp-1 model dimer, both A and B molecules superpose with the high RMSD of their Cα atoms equal to 1.21 Å, while structural discrepancies between the PR-10 group proteins are included in the range from 1.64 even to 2.76 Å [43]. Calculated in *gesamt*, the Cα RMSD between the experimental model and our Hyp-1 average low-energy conformation reaches 2.38 Å, resulting from the general flexibility of PR-10 fold and the medium size of the protein. The main secondary structure motifs were well conserved, while, as expected, most of the conformational changes occur in less stable regions of L3 and L5 loops [43]. The structural variation of Hyp-1 during simulation is also restricted by intermolecular contacts and restraints corresponding to van der Waals short-range interactions.

3.3. Actual and Experimental ADPs Comparison

In the presented ADP calculations, the simulation of the starting conformation model included several sequences: (i) 100 ps stage of the system temperature setting to 292 K, (ii) 1 ns of equilibrating simulation and (iii) 1.2 ns of the cooling of structure to 100 K followed by a sampling of the atomic positional fluctuations within the protein structure (CD and RMSF). As typically RMSDs provide a distinction between restrained and mobile parts of the molecule, we observed a reduction of flexible Cγ atoms motions under the OPLS–AA force field. This fact indicates that using the given simulation setup, we successfully generated a set of structurally diverse ensembles with high conformational heterogeneity independent from rigid-body motions or lattice defects. As the crystallographically determined Hyp-1 conformation was used as the starting model, we assumed that the resulting 64 individual protein molecules (32 dimers) are a sufficient number to present a conformational variety of Hyp-1 crystals. Positional restraints, solvent inclusion, and RMSD data acquisition from the cooled structure provided conditions similar to the experimental and possibly a high resemblance to the physical model.

Overall mean isotropic ADP for the experimental Hyp-1 model structure was equal to 27.53 Å^2, while ADPs for individual chains A and B were 23.7 Å^2 and 26.5 Å^2, respectively. The high lability and mobility of long PEG ligand chains were reflected by their increased average B factors in the range from 49 up to 71 Å^2 for the internal PEG molecules [43]. These high values result mostly from the lack of strong, directional protein–ligand interaction showing the exact composition of the PEG solution. In some cases, ligand representation in electron density maps were so uncertain that the conclusion of whether observed electron density corresponds to the full-length molecule or merely a fragment of the elongated, disordered chain was impossible. The distribution of the experimental ADPs reflects the contrast between restrained main-chain segments (Figure 3a,b) and side-chain conformational diversity (expressed by increased ADPs of Cγ atoms in Figure 3c,d). The highest values of thermal mobility of Cγ atoms were observed for side chains of Glu132, Arg93, Glu106, and Asp48 involved in hydrogen contacts to each other or with water molecules at the protein surface (Figure 3c,d).

Calculated from the simulation, the atomic RMSF maintains low values with an average ~0.2 Å within the whole simulation box and a maximum near to ~0.25 Å. Therefore, calculated ADP distribution seems to be more uniform with generally increased values of thermal factors compared to the experimental ones. Under the OPLS–AA force field and at the nanosecond timescale, the calculated mean ADPs are equal to 50.49 Å^2 (molecule A) and 47.81 Å^2 (molecule B). Their RMSD between the experimental and calculated ADPs of Cα atoms reaches values (mean $\pm$ SE) of 31.11 $\pm$ 2.63 Å^2 in chain A and 26.22 $\pm$ 2.17 Å^2 in chain B, while the total RMSD between the experimental and calculated Cα ADPs is equal to 30.51 $\pm$ 1.70 Å^2. Interestingly, the calculated Cγ ADPs seem to be more similar to the experimental distribution with their RMSD and SEs equal to 24.33 $\pm$ 2.47 Å^2 and 19.36 $\pm$ 2.00 Å^2 in relation to chains A and B, respectively. Overall, Cγ RMSD reaches 25.95 $\pm$ 1.57 Å^2. It was observed that the means and SEs of the ADPs are growing with the time of simulation (Pang, 2016)—calculated deviations between experimental and calculated ADPs of Hyp-1 protein can be explained as a result of a longer 1.2 ns trajectory.

After the re-refinement of the simulated Hyp-1 model with isotropic ADPs, diversity between residues within the structure was restored, although mean Cα ADPs for chains A and B are equal to 39.31 and 38.76 Å^2, respectively, with RMSD 20.66 $\pm$ 3.03 Å^2 (chain A) and 18.78 $\pm$ 2.56 (chain B) relative to the experimental values. This fact confirmed the systematic underestimation of X-ray experimental ADPs announced in previous studies [17]. Analogous values for Cγ atoms reach 51.25 (molecule A) and 50.88 Å^2 (molecule B), and their total RMSD to PDB entry is 17.85 $\pm$ 2.75 Å^2. We decided to use five TLS groups per each protein chain as it was implemented in the starting structure to avoid modeling whole protein as a rigid body and freezing of protein translation and rotation within the crystal structure. The distribution of ADPs within the average simulated Hyp-1 model refined with five TLS groups per each chain strongly resembles those with isotropic thermal factors. We determined the overall RMSD between isotropic and TLS-modelled ADPs at the level of 2.57 $\pm$ 0.45 Å^2 for Cα atoms and 1.68 $\pm$ 0.33 Å^2 in the case of Cγ atoms. The high flexibility of long side chains of Lys27, Glu48, Arg93, Glu102, Lys139, Glu142, and Glu149 was observed for both PDB entry and re-refined models. By introducing TLS groups and the subsequent refinement of average Hyp-1 model, simulated ADPs contained a contribution from correlated motions between neighboring atoms in the protein, which results in restricting of atomic fluctuations and a general lowering of ADPs. Contrary to

MD-based ADPs, standard thermal motions cannot always accurately reflect dynamical changes in the protein crystal structure, although we found that the simple RMSF-based calculation of ADPs is also not sufficient, probably due to the high impact of static disorder from the averaging procedure.

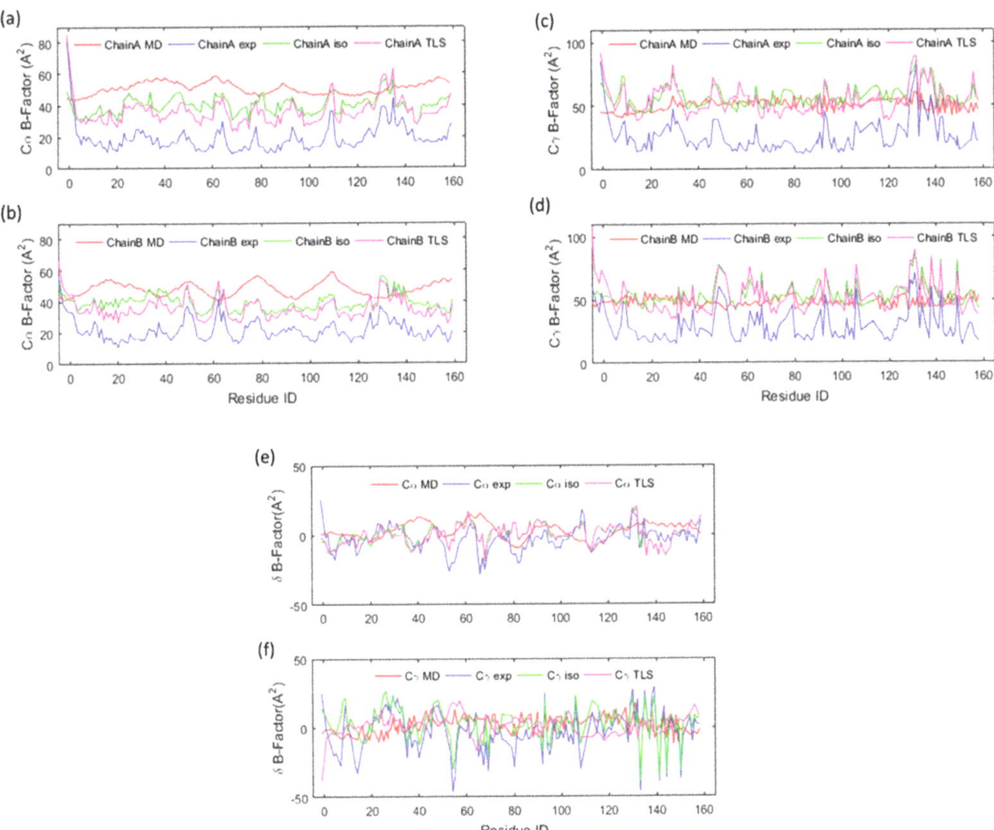

Figure 3. Plots of Hyp-1 ADPs: calculated directly from atom-positional RMSF (red line), experimental from the Protein Data Bank (PDB) entry 3IE5 (blue line), from model re-refined with isotropic thermal motions (green line) and model re-refined using translation–libration–screw model (TLS) groups (purple line). In individual panels ADPs for all CA (Cα) and CG (Cγ) atoms separated on chains A (**a**,**c**) and B (**b**,**d**) were shown. Differences in ADPs' values for both protein chains were calculated between corresponding Cα (**e**) and Cγ (**f**) atoms.

3.4. Side-Chain Angles Probability Distribution P(χ1) for Pro, Ser, Cys

The amino acid sequence of the Hyp-1 protein contains seven Pro, three Ser, and two Cys residues in a monomeric state. Proline molecules tend to adopt *endo* or *exo* conformation depending on whether displaced γ carbon is located above or below the plane formed by other α, β, δ, and N atoms [88]. Within the Hyp-1 dimer experimental model, seven Cγ-endo and seven Cγ-exo conformations of proline were expected at $\chi_1 = 30°$ and 330°, respectively. However, observed after the MD simulation, the distributions reveal interconversion between *endo* and *exo* states for Pro16, Pro64, Pro122, and Pro124 (Figure 4) marked by predicted $P(\chi_1)$ peaks at χ_1 values contrary to those resulting from experimental structure conformation.

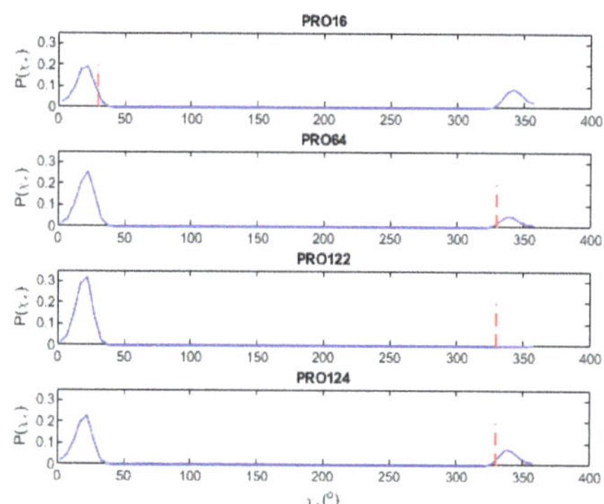

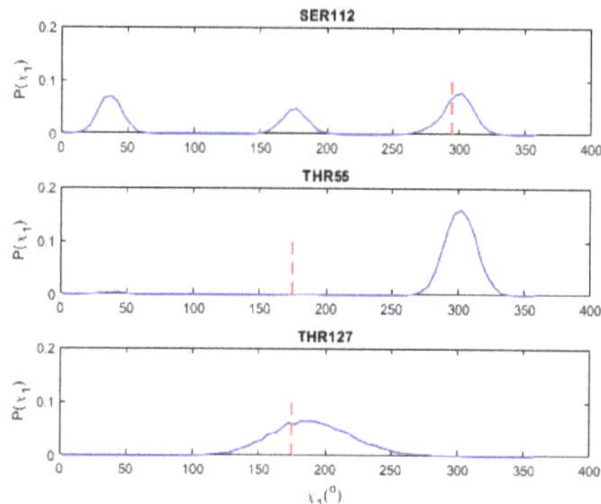

Figure 4. Side-chain torsions probability density distributions $P(\chi_1)$ (solid blue lines) for Pro16, Pro64, Pro122, and Pro124 (**left column**) and Ser112, Thr55, and Thr127 (**right column**). Values of χ_1 angles observed in the experimental Hyp-1 model were marked as dashed red lines.

The high flexibility of the pyrrolidine ring and the dynamic tendency toward rapid *endo–exo* and vice versa conversion at the Cγ atom was previously confirmed by the analysis of the ^{1}H NMR spectrum of proline in aqueous solution [88]. Although the Cγ atom in most of Pro residues exists in one of *endo–exo* conformation for both A and B protein molecules, their $P(\chi_1)$ distributions show that conversion to energetically preferred $\chi_1 = 30°$ (*endo* form) is incomplete. Only for Pro122, a peak by $\chi_1 = 330°$ is nearly absent, indicating a low affinity for this conformation. Due to the location of most proline residues in loop areas of the Hyp-1 structure (except for Pro16 from short helix α1), they possess high conformational freedom facilitating the transformation to more stable *endo* conformation. For serine and cysteine molecules, three highly probable side-chain conformations can be expected at $\chi_1 = 62°$ (**p**), $\chi_1 = 183°$ (**t** form) and $\chi_1 = 295°$ (**m**). We did not find any discrepancies between experimental and predicted Cys rotameric form, which indicates the convergence of knowledge-based and physics-based approaches. As shown in Figure 4, a disproportion between experimental and predicted rotameric form was found in Ser112 from the L8 loop, where highly probable **p** conformation is expected at $\chi_1 = 295°$. Calculated $P(\chi_1)$ distribution reveals three peaks at 295°, 183°, and 30°, suggesting the creation of different energy minima for each conformation. The deviation of peak value at $\chi_1 = 30°$ from the preferred 62° value suggests some geometrical distortion from the experimentally defined model or energetic affinity to one of the non-rotameric states.

3.5. Val and Thr

According to the Rotamer Library, three mostly preferred Val and Thr side-chain conformations can be expected near $\chi_1 = 60°$, 175° and 300°. The native protein sequence of Hyp-1 contains 18 valine residues and 16 of them, regardless of their location in secondary structure, adopt dominant **t** rotameric state from the Penultimate Library. Surprisingly, we noted differences between the experimental and simulated side-chain conformations for 17 of 18 Hyp-1 valine residues, so their predicted $P(\chi_1)$ distribution almost always has a peak at nearly 300° respective to the second most probable rotamer **m**. The occurrence of lesser-populated rotameric states parallel to more extended or rarer conformations indicates dynamic conformational changes within the model structure during MD simulation. This conformational transformation towards lower energy states was expressed by a probability shift from the **t** to **m** rotamer and the resulting switch in χ_1 preference. In the case of nine Thr molecules from the Hyp-1 sequence, eight residues have a different simulated rotamer with probability shifted significantly from preferred **p** (49% independent probability) and **m** (43% of probability) to the least populated of the three top Thr conformations **t** form. As recommended by The International Union of Pure and Applied Chemistry IUPAC, different definitions of χ_1 in side-chain methyl group, $\chi_1 = 175°$ in Val and $\chi_1 = 295°$ in Thr are equivalent. For β-branched Thr55, we noticed the reversion between two dominating rotameric states as it was for many Val residues (Figure 4). When it comes to Thr127 in a less ordered loop area, a decrease in the **t** probability and subsequent shift toward preferred **m** rotamer was observed (Figure 4). More details on the side-chain torsions probability density distributions for Val residues can be found in Figures S1–S3 (see Supplementary Materials).

3.6. Leu, Phe, Tyr, His

Preferred side-chain conformations for Leu and aromatic Phe, Tyr, His residues were modeled using dihedral angles bimodal P(χ_1, χ_2) distributions. Although the Hyp-1 native sequence contains 10 hydrophobic Leu residues, only two of them were marked to have different rotameric forms in experimental and simulated structure models. For β-branched Leu86, χ_1 and χ_2 are expected near to 175° and 65°, respectively, while in the calculated distribution "vertical" χ_2 transition to ~180° was noted resulting in rare side-chain conformations (Figure 5a). Because of the known fact that sparsely populated rotamers coincide with higher-energy parts of energy landscapes [89] and knowing positions of potential catalytic active sites in structure, it is particularly interesting, because Leu86 was introduced into Hyp-1 as a substituent for original Ile86 and therefore considered as one of the mutations blocking the enzymatic activity of Hyp-1 in hypericin synthesis [43].

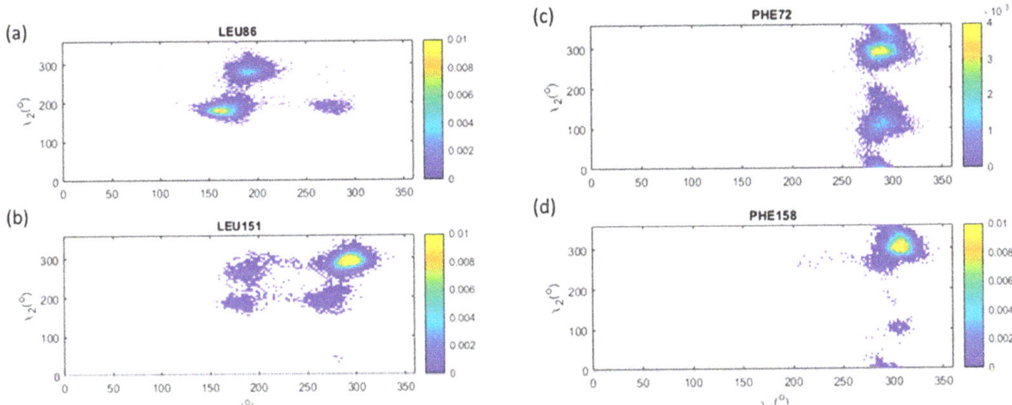

Figure 5. Color map of side-chain torsions probability density distributions P(χ1, χ2) for Leu86, Leu151 (**a**,**b**) and aromatic residues Phe72, Phe158 (**c**,**d**). The probability values within each of the 5° × 5° bins increase from deep blue to yellow according to the color map.

Figure 5b presents that another rotamer outlier from the experimental model was observed for Leu151 in the helix α3 region, where the calculated χ_1 distribution is centered around the rotameric value of 295° with a non-rotameric peak of χ_2 around 300°. We typically suspect at least one non-rotameric torsion in many side chains, especially those long and characterized by many torsions, but it remains an inquisitive observation in the case of the shorter, aliphatic Leu molecule. This side-chain conformation anomaly suggests the high energy perturbations in the carboxylic terminal group of Leu151, evidenced by its increased C_γ ADP value (Figure 3c,d). MD simulation studies showed considerable freedom of the Leu151 side-chain conformation suggesting its possible role in substrate recognition. Indeed, further analysis of the Hyp-1/ANS complex revealed that typically very non-reactive, hydrophobic Leu151 can be involved in ligand binding. We did not find any differences between the experimental and simulated rotameric forms in the dipeptide mimic of Leu, Ile.

Aromatic Phe amino acid in the Hyp-1 structure is represented by eight residues in each protein molecule, most of them buried in a hydrophobic internal cavity. We compared the experimental rotamers with their simulated forms revealing differences at Phe72 and

Phe158 (Figure 5c,d). The first of them was initially modeled into electron density with a double side-chain conformation, **m-30**, and less occupied **m-85**. Both experimental and calculated χ_1 are close to 295°, whereas the observed χ_2 distribution has a maximum of around 270°. The second densely populated χ_2 region occupies the area in the proximity of 90°, resulting from a 180° ring inversion and the subsequent creation of two indistinguishable side-chain conformations. Contrarily, the χ_2 pair from the experimental model should show peaks around 330° and 150°, even if calculated P(χ_2) distribution clearly reveals minor probabilities in these regions. A subtle difference was observed for the simulation of C-terminal Phe158 residue with a preference of χ_2 torsion distribution to ~300° value. The previous exploration of the Phe conformation richness in protein structures revealed that χ_1 = 300° could be strongly limited by steric clashes between the aromatic ring and the carbonyl group of adjoining residues, especially within α-helical regions [79]. However, we found this value most representative for five of eight Phe amino acids in each Hyp-1 chain. Moreover, we discovered using a physics-based approach that the **m-85** rotamer is the energetically favorable conformation of the Phe72 side chain, which was unclear during refinement with arbitrarily selected and incorrectly refined occupancies of different side chains.

Among six His residues from the Hyp-1 protein sequence, our hard-sphere calculations revealed two main derogations from experimental side-chain stereochemistry (Figure 6a,b). For His63, strong peaks at χ_1 = 180° and χ_2 = 260° correspond to **t-80** rotamer from the structure model. However, the large patch in P(χ_1, χ_2) distribution was observed around χ_1 = 295° and χ_2 = 290° respective to dominant **m-70** state populated 29% of the time, although this conformation was absent in the initial dataset. A second noticeable sample is pocket-hidden His70 with the experimental second most preferred rotamer **m80** (χ_1 = 295°, χ_2 = 80°). The calculated distribution indicates rather on less observed **m170** form (7% of probability) in agreement with the observation of χ_1 and χ_2 peaks near 295° and 165°.

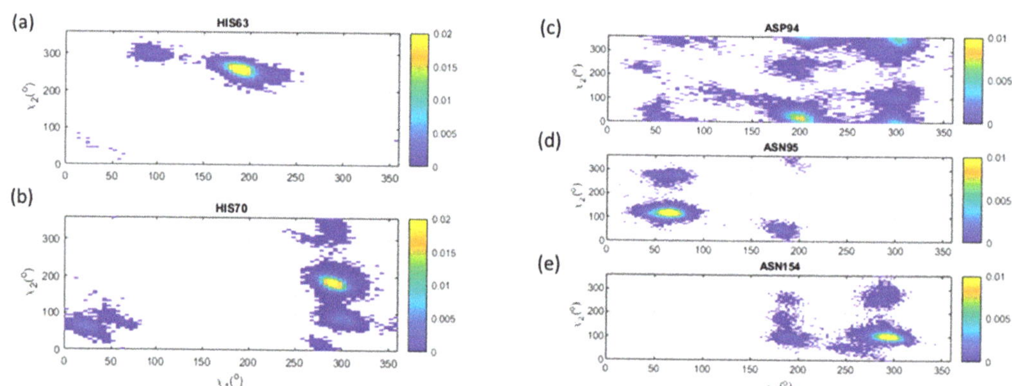

Figure 6. Color map of the side-chain torsions probability density distributions P(χ_1, χ_2) for aromatic residues His63, His70 (**a**,**b**), and Asp94, Asn95, and Asn154 (**c**–**e**). The probability values within each of the 5° × 5° bins increase from deep blue to yellow according to the color map.

We observed no differences between the experimental and calculated rotameric form of Tyr with analogous to Phe aromatic side chain. Despite the structural resemblance between both amino acids and the ease of their mutual substitution, we found Tyr to be a generally more stable residue in the Hyp-1 protein structure with low sensibility on energy-driven conformational changes.

3.7. *Asp, Asn*

Although the native sequence of Hyp-1 contains 6 Asp residues, we found only one disagreement between experimental and simulated Asp94 rotamers. As in the PDB-derived Hyp-1 model, we might expect two highly populated clusters around $\chi_1 = 290°$ and $\chi_2 = 345°$ corresponding to the most populated **m-20** conformation. On the contrary, we found in Figure 6c a significant blur of allowed side-chain torsions in the calculated $P(\chi_1, \chi_2)$ distribution of Asp94, where strong peaks were found at $\chi_1 \sim 200°$ and χ_2 near to 0°. Troubles with assigning the central value and standard deviation to the suggested conformation mark a high level of conformational heterogeneity of the Asp94 residue confirmed by the higher ADP of its Cγ atom (Figure 3c,d). Furthermore, our studies imply that the transition to rare rotameric states is energetically preferred. Similar conclusions emerge from the analysis of $P(\chi_1, \chi_2)$ distribution in Asn95 and Asn154 (Figure 6d,e). Both of them exist in uncommon **p-10** and **m-80** conformations that should have a frequent population near $\chi_1 = 62°$ and 295°, respectively. Our $P(\chi_1, \chi_2)$ calculations clearly confirm that despite the well preserved χ_1 value, the more mobile outermost χ_2 torsion determines the rotamer form and plays a crucial role in the examination of preferred Asn conformations. For Asn95, the χ_2 peak near 90° favors more popular **p30** conformation and in the case of Asn154, the rare **m120** rotamer is suspected because of the $P(\chi_2)$ area around 100°. The known problem with Asn conformers is that its $P(\chi_2)$ distribution is typically broad and ambiguous, and the strict value of χ_2 can differ by 180° due to the possible flip of identical side-chain amide N and O atoms [90]. As an effect, symmetric sparsely populated regions with χ_2 changing by 180° are observed on our calculated $P(\chi_1, \chi_2)$ distribution. We overcome this problem with the poor clustering of χ torsions and the less meaningful analysis of allowed Asn conformations. Our $P(\chi_1, \chi_2)$ shows separated local clusters of χ_2 values clearly determining the preferred Asn rotamer. We observed a lack of these sharp, clear clusters for Asp94, which we assigned to large mobility of partially disordered side chains and high-energy perturbations within the terminal carboxylic group.

3.8. Glu, Gln

We investigated experimental rotamers with the full density distribution of χ_1, χ_2, χ_3 dihedral angles for each of 19 Glu and three Gln residues in a single Hyp-1 chain. Deviations were noticed for six polar, surface Glu amino acids modeled in most common rotamers, **mt-10** (33% of probability), or **tt0** (24% of all Glu) in accordance with the Penultimate Rotamer Library. Under the simulation regime, we found the tendency of these residues to adopt rare side-chain conformations, e.g., **tm-20** and **tp10**. The broadening of individual χ distributions (especially for terminal χ_3 between sp^3 and sp^2 hybridized atoms) complicates the determination of the rotameric state (Figure 7a,b). As in the case of the experimental Hyp-1 structure model, simulated Glu conformations are mainly stabilized by H-bonds between solvent and side-chain amide or carboxyl group. For α3-helical Glu142 by cavity entrance EB (Figure 7a), pairing with the protonated amino group of adjacent Asn134 may occur and impose the characteristic codependency of χ dihedral angles to preserve this preferred interaction. A comparison between the experimental and simulated Hyp-1 models have shown rotamer outliers on Gln35 and Gln146 (Figure 7c,d).

On the basis of the experimental Hyp-1 structure, strong peaks at $P(\chi_1, \chi_2, \chi_3)$ distribution for the most common rotamer **mt-30** are expected near $\chi_1 = 290°$, $\chi_2 = 180°$ and $\chi_3 = 330°$. The complementary rotameric form was stabilized by mutual hydrogen interactions of carboxyl and amino groups. This strengthening pattern was broken under simulation conditions, in which steric restraints and greater solvent accessibility reversed Gln35 and Gln146 side-chain conformations. In our simulated model, less common rotamers **tt0**, **mm-40**, and **mm100** are favored and stabilized by interactions with surrounding solvent molecules. Preference to adopt rare rotamers in long side chains of Glu and Gln arises from their susceptibility to dynamic changes, surface exposition, and the possible lack of strong intramolecular interactions [39]. Due to this, it is usually recommended to model such side chains using only one or several most common rotamers [38]. Analyzing the rotamericity of Glu/Gln side-chain conformations, we found this approach to be insufficient, because some certain, energetically favored conformations can be omitted. As a result, we noticed some inexplicable ADP peaks at the Cγ atom of Glu132 from the experimental Hyp-1 structure

(Figure 3c,d), although the residue was modeled in dominant rotameric form. The swap to supposedly less popular conformation minimizes the thermal motions of Glu132 indicating lower energy conformation.

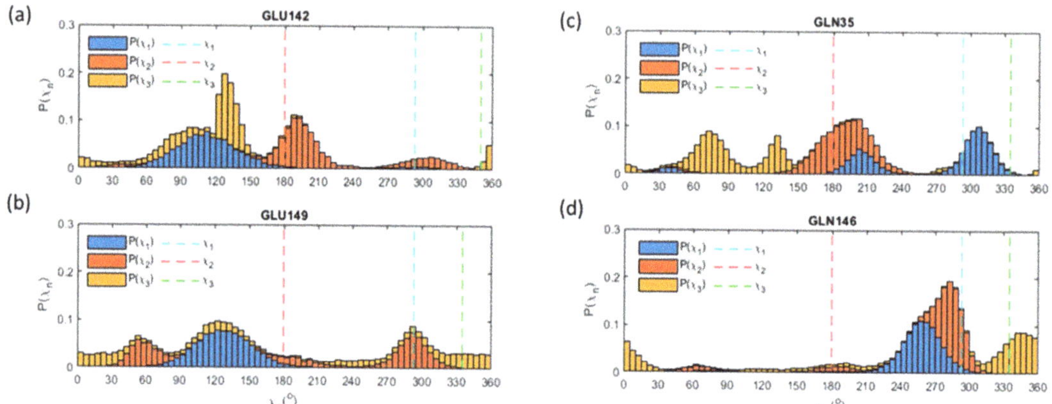

Figure 7. Stacked bar graph of side-chain χ_1, χ_2, χ_3 torsions probability distributions for Glu142, Glu149 (**a**,**b**), and Gln35, Gln146 (**c**,**d**). The probability values within 5° × 5° bins for separate torsion angles were marked as blue, orange, and yellow, respectively. Appropriate values of χ_1, χ_2, χ_3 angles from the experimental Hyp-1 model were presented sequentially as light blue, orange, and light green dashed lines.

3.9. Met

Amino-acid sequence of Hyp-1 contains only two Met residues, N-terminal Met1, and β-branched Met68, in each of the protein molecules within the dimer. According to the experimental rotamers, one might expect $P(\chi_1, \chi_2, \chi_3)$ distribution peaks at $\chi_1 = 293°$, $\chi_2 = 180°$ and $\chi_3 = 180°$ or $\chi_3 = 285°$. The value of χ_3 clearly determines the available rotamer, as other dihedral angles were the same for both conformations, **mtt**, and **mtm**. They belong to less populated rotamer states, with 8 and 11% of appearance according to Penultimate Library. In our studies, both simulated Met rotamers diverge from those experimental, mainly towards more popular **mtp** conformation indicated by χ_3 near 75° (Figure 8a,b). Our physics-based approach confirmed well the preservation of two χ_1, χ_2 angles between sp^3 atoms attached to the δ atom [91]. The main difference lies in the terminal χ_3 due to the elongated C–S bond and accompanying a low rotational barrier in comparison to other all-carbon tetrahedral torsions. High conformational freedom of Met residues could be also a result of a lack of directional interactions stabilizing the C-terminal part of the side chain. In our simulated model, the choice of rotamer is dictated by bonds with penetrating solvent as well as strong repulsing restraints imposed during a simulation. The supposed conformation of Met68 seems to be particularly important, as this residue was involved in ligand binding and the forming of Hyp-1 protein complexes.

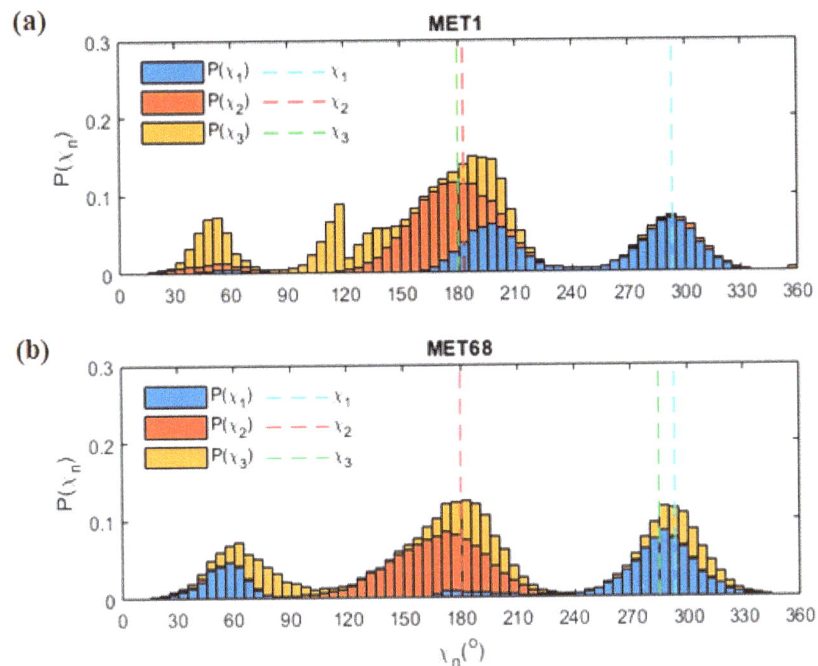

Figure 8. Stacked bar graph of side-chain χ_1, χ_2, χ_3 torsions probability distributions for Met1 (**a**), and Met68 (**b**). The probability values within 5° × 5° bins for separate torsion angles were marked as blue, orange, and yellow, respectively. Appropriate values of χ_1, χ_2, χ_3 angles from the experimental Hyp-1 model were presented sequentially as light blue, orange, and light green dashed lines.

3.10. Lys, Arg

Long side chains of Arg and Lys characterized by four dihedral angles can occur in many combinations. Among 16 Lys residues in a single Hyp-1 molecule, all of them adopt rare rotameric states with Lys123 **mttt** (20% of samples) as the most popular one. In eight cases (Lys8, 21, 33, 40, 83, 113, 138 and 145), the deviations between experimental and simulated rotamers were observed, each of them suggesting a more populated conformation (Figure 9a,b, Figures S8–S10 (see Supplementary Materials)).

The disordered structure of Lys8 prevents bonding with the PEG501 molecule, while in our studies, the favored **mttt** conformation provides optimal hydrogen interactions with both the surrounding ligand and solvent. In the Hyp-1 structure, Lys side chains tend to form partially ordered supramolecular crown ethers adducts with PEG ligands, first observed for Lys33 in molecule A. Specific interaction with PEG ligands restricts rare conformations of Lys33, while our simulation suggests that conformational changes of the Lys33A side chain can improve the hydrogen interaction with the lysine terminal amino group. In chain B, a different orientation of Lys33 blocks the possibility of effective PEG binding and imposes another conformation of the ligand. Positively charged Lys residues can be easily replaced by Arg, while three of them exist in rare **mtp180** and **ptm-85** rotamers in each protein chain. Our calculated 4-dimensional probability distribution of torsion angles reveals conformational changes in Arg27 and Arg93. P(χ_1, χ_2, χ_3, χ_4) distribution of Arg27 side-chain angles clearly exhibits the most populated sharp clusters referring to each torsion. Although Arg27 was modeled in double conformation in an experimental Hyp-1 structure as a result of stabilizing interaction absence, during the simulation we found **ttp-105** to be the most preferred rotamer. Therefore, H-bonds with an adjacent hydroxyl group of Tyr85 and backbone Leu24 carboxyl group are responsible for the stabilization of simulated rotamer and its anchoring within the protein interior. Within the probability distribution of Arg93 side-chain torsions, we found broad conformational regions of terminal χ_3 and χ_4 angles (Figure 9d). Due to this reason, the conformational diversity of Arg93 is expressed as a mixture of differently populated rotamers. MD simulation can indicate some beneficial regions in the energy landscape of structure, especially for **mtt85** rotamer, in which interactions of the guanidinium group of Arg93 side chain by EB entrance define the internal pocket availability for ligands.

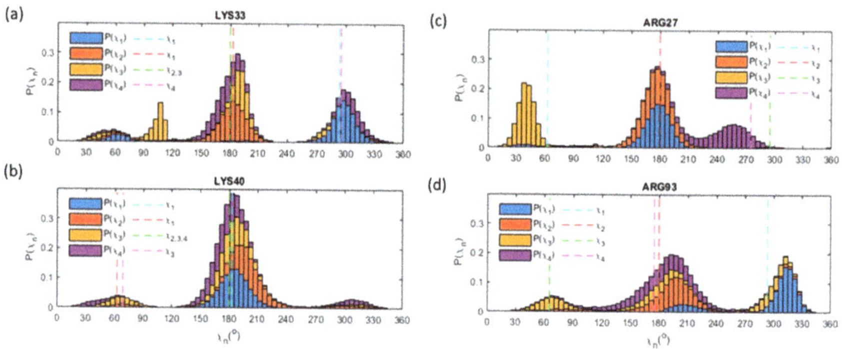

Figure 9. Stacked bar graph of side-chain χ_1, χ_2, χ_3, χ_4 torsions probability distributions for Lys33 (**a**), Lys40 (**b**), Arg27 (**c**) and Arg93 (**d**). Each probability distribution of χ_i was normalized so that $\int P(\chi_i)dx_i = 1$. The probability values within 5° × 5° bins for separate torsion angles were marked as blue, orange, yellow, and purple, respectively. Appropriate values of χ_1, χ_2, χ_3 angles from the experimental Hyp-1 model were presented sequentially as light blue, orange, and light green dashed lines. If one χ_i value was common for multiple torsion angles in the side chain, we marked them as a single line with a proper index to prevent line overlap and confusion during figure interpretation.

4. Conclusions

In the presented work, we compared the experimental and deposited in the PDB database protein structure of Hyp-1 with the results of the molecular dynamics simulations. We tried to use a physics-based approach as complementary to the knowledge-based method of protein structure validation. Our results confirmed that multistep MD simulation can successfully maintain the secondary structure of the protein and its characteristic fold without disturbances of main secondary structure motifs, e.g., α helices, and β strands. Calculated Cα RMSD equal to 2.38 Å results mainly from the high conformational freedom of protein side chains and is comparable with the results of ns-scale protein simulation mentioned in other papers. Direct observation of atomic trajectories during simulation allows calculating individual RMSD and ADPs, which are typically >10 Å^2 higher than experimental ones. Our averaged Hyp-1 model was re-refined with isotropic ADPs and using the TLS model, which restored diversity between ADPs within the structure due to more defined restraints imposed on the backbone and side chains and inclusion of correlated atomic motions in opposition to pure dynamic changes during simulation. We found our result consistent with other MD simulations of protein models, in which experimental ADPs appear to be systematically understated. Comparing average structure factors, a high level of similarity between those calculated and simulated was observed. We used the MD method as a tool for the verification of experimental Hyp-1 rotamers because at least part of them was arbitrarily modeled into electron density indicated by increased ADPs at their Cγ atoms and simulation seems to provide more rigorous insight into dynamical properties of the system. Based on the normalized probability density distribution of χ_i torsion angles from the optimal 1 ns simulation, we proposed energetically preferred rotameric forms at 292 K differentt to thoset resulting from the refinement. We explained how their existence was connected with the structural properties of the Hyp-1 protein, especially in the context of possible biologically active ligand binding. A complete comparison of rotamer outliers within Hyp-1 experimental and simulated structures, partially with a calculated probability of each angle samples within 40° × 40° boxes, is provided in Tables S1 and S2 of Supplementary Content (see Supplementary Materials). The method of MD simulation appears to be particularly useful for the examination and validation of post-refinement protein structures complementary to the currently used knowledge-based techniques. A near-experimental approach is a key factor in a proper understanding of the functional role of the protein. Using MD-based sampling simulations, some structural irregularities in bonds, angles, or rotameric states can be corrected with sufficient accuracy to monitor the biological activity of the protein. It is crucial in the case of seemingly uncomplicated protein structures such as Hyp-1 with potentially high potential in medicine and affinity to various ligand binding. We affirmed MD methods as a useful tool in the verification of experimental protein models and the explanation of "blank spaces" in poorly refined regions of electron density maps. In the future, we would like to develop this approach on a larger scale and use it to explain other structural ambiguities, mainly in larger Hyp-1 complexes with different ligands and their bizarrely modulated phases.

Supplementary Materials: The following are available online at https://www.mdpi.com/2073-4352/11/1/43/s1.

Author Contributions: Conceptualization, J.S. and R.S.; methodology, T.K.; software, T.K.; validation, T.K., J.S., R.S. and I.B.; formal analysis, R.S. and J.W.; investigation, J.S.; resources, J.S. and T.K.; data curation, T.K.; writing—original draft preparation, J.S.; writing—review and editing, R.S. and T.K.; visualization, J.S.; supervision, J.W.; project administration, J.W.; funding acquisition, J.W. All authors have read and agreed to the published version of the manuscript.

Funding: Joanna Smietanska is partly supported by an EU Project POWR.03.02.00-00-I004/16. Authors acknowledge financial support from Polish National Science Center under grant no. 2019/33/B/ST3/02063.

Informed Consent Statement: Informed consent was obtained from all subjects involved in the study.

Data Availability Statement: The data presented in this study are available in Supplementary Material.

Acknowledgments: We would like to show our gratitude to the Mariusz Jaskolski from Center for Biocrystallographic Research in Poznan, Poland, for many discussions and his insightful comments that greatly improved the manuscript.

Conflicts of Interest: The authors declare no conflict of interest.

References

1. McCammon, J.A.; Gelin, B.R.; Karplus, M. Dynamics of folded proteins. *Nat. Cell Biol.* **1977**, *267*, 585–590. [CrossRef]
2. Gaalswyk, K.; Muniyat, M.I.; Maccallum, J.L. The emerging role of physical modeling in the future of structure determination. *Curr. Opin. Struct. Biol.* **2018**, *49*, 145–153. [CrossRef]
3. Patodia, S.; Bagaria, A.; Chopra, D. Molecular Dynamics Simulation of Proteins: A Brief Overview. *J. Chem. Phys. Biophys.* **2014**, *4*. [CrossRef]
4. Karplus, M.; Kuriyan, J. Molecular dynamics and protein function. *Proc. Natl. Acad. Sci. USA* **2005**, *102*, 6679–6685. [CrossRef]
5. Shaw, D.E.; Maragakis, P.; Lindorff-Larsen, K.; Piana, S.; Dror, R.O.; Eastwood, M.P.; Bank, J.A.; Jumper, J.M.; Salmon, J.K.; Shan, Y.; et al. Atomic-Level Characterization of the Structural Dynamics of Proteins. *Science* **2010**, *330*, 341–346. [CrossRef]
6. Freddolino, P.L.; Arkhipov, A.S.; Larson, S.B.; McPherson, A.; Schulten, K. Molecular Dynamics Simulations of the Complete Satellite Tobacco Mosaic Virus. *Structure* **2006**, *14*, 437–449. [CrossRef]
7. Perez, A.; Morrone, J.A.; Simmerling, C.; Dill, K.A. Advances in free-energy-based simulations of protein folding and ligand binding. *Curr. Opin. Struct. Biol.* **2016**, *36*, 25–31. [CrossRef]
8. Allen, F.; Almasi, G.; Andreoni, W.; Beece, D.; Berne, B.J.; Bright, A.; Brunheroto, J.; Cascaval, C.; Castanos, J.; Coteus, P.; et al. Blue Gene: A vision for protein science using a petaflop supercomputer. *IBM Syst. J.* **2001**, *40*, 310–327. [CrossRef]
9. Brooks, B.R.; Bruccoleri, R.E.; Olafson, B.D.; States, D.J.; Swaminathan, S.; Karplus, M. CHARMM: A program for macromolecular energy, minimization, and dynamics calculations. *J. Comput. Chem.* **1983**, *4*, 187–217. [CrossRef]
10. Phillips, J.C.; Braun, R.; Wang, W.; Gumbart, J.; Tajkhorshid, E.; Villa, E.; Chipot, C.; Skeel, R.D.; Kalé, L.; Schulten, K. Scalable molecular dynamics with NAMD. *J. Comput. Chem.* **2005**, *26*, 1781–1802. [CrossRef] [PubMed]
11. Abraham, M.J.; Murtola, T.; Schulz, R.; Pall, S.; Smith, J.C.; Hess, B.; Lindahl, E. GROMACS: High performance molecular simulations through multi-level parallelism from laptops to supercomputers. *SoftwareX* **2015**, *1*, 19–25. [CrossRef]

12. Cornell, W.D.; Cieplak, P.; Bayly, C.I.; Gould, I.R.; Merz, K.M.; Ferguson, D.M.; Spellmeyer, D.C.; Fox, T.; Caldwell, J.W.; Kollman, P.A. A Second Generation Force Field for the Simulation of Proteins, Nucleic Acids, and Organic Molecules. *J. Am. Chem. Soc.* **1995**, *117*, 5179–5197. [CrossRef]
13. Lee, M.R.; Tsai, J.; Baker, D.; Kollman, P.A. Molecular dynamics in the endgame of protein structure prediction. *J. Mol. Biol.* **2001**, *313*, 417–430. [CrossRef] [PubMed]
14. Karplus, M.; McCammon, J.A. Molecular dynamics simulations of biomolecules. *Nat. Struct. Biol.* **2002**, *9*, 646–652. [CrossRef]
15. Dror, R.O.; Dirks, R.M.; Grossman, J.; Xu, H.; Shaw, D.E. Biomolecular Simulation: A Computational Microscope for Molecular Biology. *Annu. Rev. Biophys.* **2012**, *41*, 429–452. [CrossRef]
16. Chen, J.; Brooks, C.L. Can molecular dynamics simulations provide high-resolution refinement of protein structure? *Proteins Struct. Funct. Bioinform.* **2007**, *67*, 922–930. [CrossRef]
17. Kuzmanic, A.; Pannu, N.S.; Zagrovic, B. X-ray refinement significantly underestimates the level of microscopic heterogeneity in biomolecular crystals. *Nat. Commun.* **2014**, *5*. [CrossRef]
18. Piana, S. How robust are protein folding simulations with respect to force field parameterization? *Biophys. J.* **2011**, *101*, 1015. [CrossRef]
19. Izmailov, S.A.; Podkorytov, I.S.; Skrynnikov, N.R. Simple MD-based model for oxidative folding of peptides and proteins. *Sci. Rep.* **2017**, *7*, 9293. [CrossRef]
20. Ensign, D.L.; Kasson, P.M.; Pande, V.S. Heterogeneity Even at the Speed Limit of Folding: Large-scale Molecular Dynamics Study of a Fast-folding Variant of the Villin Headpiece. *J. Mol. Biol.* **2007**, *374*, 806–816. [CrossRef]
21. Stumpff-Kane, A.W.; Maksimiak, K.; Lee, M.S.; Feig, M. Sampling of near-native protein conformations during protein structure refinement using a coarse-grained model, normal modes, and molecular dynamics simulations. *Proteins Struct. Funct. Bioinform.* **2007**, *70*, 1345–1356. [CrossRef] [PubMed]
22. Wereszczynski, J.; McCammon, J.A. Statistical mechanics and molecular dynamics in evaluating thermodynamic properties of biomolecular recognition. *Q. Rev. Biophys.* **2011**, *45*, 1–25. [CrossRef] [PubMed]
23. Adiyaman, R.; McGuffin, L.J. Methods for the Refinement of Protein Structure 3D Models. *Int. J. Mol. Sci.* **2019**, *20*, 2301. [CrossRef] [PubMed]
24. Wong-Ekkabut, J.; Karttunen, M. The good, the bad and the user in soft matter simulations. *Biochim. Biophys. Acta Biomembr.* **2016**, *1858*, 2529–2538. [CrossRef]
25. Bernardi, R.C.; Melo, M.C.R.; Schulten, K. Enhanced sampling techniques in molecular dynamics simulation of biological systems. *Biochim. Biophys. Acta* **2015**, *1850*, 872–877. [CrossRef] [PubMed]
26. Feig, M.; Mirjalili, V. Protein structure refinement via molecular-dynamics simulations: What works and what does not? *Proteins Struct. Funct. Bioinform.* **2015**, *84*, 282–292. [CrossRef]
27. Pang, Y.-P. Use of multiple picosecond high-mass molecular dynamics simulations to predict crystallographic B-factors of folded globular proteins. *Heliyon* **2016**, *2*. [CrossRef]
28. Masmaliyeva, R.C.; Murshudov, G.N. Analysis and validation of macromolecular B values. *Acta Crystallogr. Sect. D Struct. Biol.* **2019**, *75*, 505–518. [CrossRef]
29. Caldararu, O.; Kumar, R.; Oksanen, E.; Logan, D.T.; Ryde, U. Are crystallographic B-factors suitable for calculating protein conformational entropy? *Phys. Chem. Chem. Phys.* **2019**, *21*, 18149–18160. [CrossRef]
30. Parthasarathy, S.; Murthy, M. Analysis of temperature factor distribution in high-resolution protein structures. *Protein Sci.* **2008**, *6*, 2561–2567. [CrossRef]
31. Carugo, O.; Argos, P. Correlation between side chain mobility and conformation in protein structures. *Protein Eng. Des. Sel.* **1997**, *10*, 777–787. [CrossRef] [PubMed]
32. Meinhold, L.; Smith, J.C. Fluctuations and Correlations in Crystalline Protein Dynamics: A Simulation Analysis of Staphylococcal Nuclease. *Biophys. J.* **2005**, *88*, 2554–2563. [CrossRef] [PubMed]
33. Kuriyan, J.; Weis, W.I. Rigid protein motion as a model for crystallographic temperature factors. *Proc. Natl. Acad. Sci. USA* **1991**, *88*, 2773–2777. [CrossRef]
34. Soheilifard, R.; Makarov, D.E.; Rodin, G.J. Critical evaluation of simple network models of protein dynamics and their comparison with crystallographic B-factors. *Phys. Biol.* **2008**, *5*, 026008. [CrossRef] [PubMed]
35. Winn, M.D.; Isupovb, M.N.; Murshudov, G.N. Use of TLS parameters to model anisotropic displacements in macromolecular refinement. *Acta Cryst.* **2001**, *57*, 122–133. [CrossRef] [PubMed]
36. Carugo, O. How large B-factors can be in protein crystal structures. *BMC Bioinform.* **2018**, *19*. [CrossRef] [PubMed]
37. Kato, K.; Nakayoshi, T.; Fukuyoshi, S.; Kurimoto, E.; Oda, A. Validation of Molecular Dynamics Simulations for Prediction of Three-Dimensional Structures of Small Proteins. *Molecules* **2017**, *22*, 1716. [CrossRef]
38. Hintze, B.J.; Lewis, S.M.; Richardson, J.S.; Richardson, D.C. Molprobity's Ultimate Rotamer-Library Distributions for Model Validation. *Proteins* **2016**, *84*, 1177–1189. [CrossRef]
39. Towse, C.-L.; Rysavy, S.J.; Vulovic, I.M.; Daggett, V. New Dynamic Rotamer Libraries: Data-Driven Analysis of Side-Chain Conformational Propensities. *Structure* **2016**, *24*, 187–199. [CrossRef]
40. Harder, T.; Boomsma, W.; Paluszewski, M.; Frellsen, J.; Johansson, K.E.; Hamelryck, T. Beyond rotamers: A generative, probabilistic model of side chains in proteins. *BMC Bioinform.* **2010**, *11*. [CrossRef]

41. Haddad, J.; Vojtech, A.; Heger, Z. Rotamer Dynamics: Analysis of Rotamers in Molecular Dynamics Simulations of Proteins. *Biophys. J.* **2019**, *116*, 2062–2072. [CrossRef] [PubMed]
42. Butterweck, V. Mechanism of action of St. John's wort in depression: What is known? *CNS Drugs* **2003**, *17*, 539–562. [CrossRef] [PubMed]
43. Michalska, K.; Fernandes, H.; Sikorski, M.M.; Jaskolski, M. Crystal structure of Hyp-1, a St. John's wort protein implicated in the biosynthesis of hypericin. *J. Struct. Biol.* **2010**, *169*, 161–171. [CrossRef] [PubMed]
44. Sliwiak, J.; Dauter, Z.; Jaskolski, M. Hyp-1 protein from St John's wort as a PR-10 protein. *Biotechnology* **2013**, *1*, 47–50. [CrossRef]
45. Sliwiak, J.; Dauter, Z.; Jaskolski, M. Crystal Structure of Hyp-1, a Hypericum perforatum PR-10 Protein, in Complex with Melatonin. *Front. Plant Sci.* **2016**, *7*, 668. [CrossRef]
46. Sliwiak, J.; Dauter, Z.; Kowiel, M.; McCoy, A.; Read, R.J.; Jaskolski, M. ANS complex of St. John's wort PR-10 protein with 28 copies in the asymmetric unit: A fiendish combination of pseudosymmetry with tetartohedral twinning. *Acta Cryst.* **2015**, *D71*, 829–843. [CrossRef]
47. Berendsen, H.; Van Der Spoel, D.; Van Drunen, R. GROMACS: A message-passing parallel molecular dynamics implementation. *Comput. Phys. Commun.* **1995**, *91*, 43–56. [CrossRef]
48. Lindahl, E.; Hess, B.; Van Der Spoel, D. GROMACS 3.0: A package for molecular simulation and trajectory analysis. *J. Mol. Model.* **2001**, *7*, 306–317. [CrossRef]
49. Van Der Spoel, D.; Lindahl, E.; Hess, B.; Groenhof, G.; Mark, A.E.; Berendsen, H.J.C. GROMACS: Fast, flexible, and free. *J. Comput. Chem.* **2005**, *26*, 1701–1718. [CrossRef]
50. Hess, B.; Kutzner, C.; Van Der Spoel, D.; Lindahl, E. GROMACS 4: Algorithms for Highly Efficient, Load-Balanced, and Scalable Molecular Simulation. *J. Chem. Theory Comput.* **2008**, *4*, 435–447. [CrossRef]
51. Pronk, S.; Páll, S.; Schulz, R.; Larsson, P.; Bjelkmar, P.; Apostolov, R.; Shirts, M.R.; Smith, J.C.; Kasson, P.M.; Van Der Spoel, D.; et al. GROMACS 4.5: A high-throughput and highly parallel open source molecular simulation toolkit. *Bioinformatics* **2013**, *29*, 845–854. [CrossRef] [PubMed]
52. Pall, S.; Abraham, M.J.; Kutzner, C.; Hess, B.; Lindahl, E. Tackling Exascale Software Challenges in Molecular Dynamics Simulations with GROMACS. In *Solving Software Challenges for Exascale*; Markidis, S., Laure, E., Eds.; Springer: Cham, Switzerland, 2015; pp. 3–27.
53. Abraham, M.J.; van der Spoel, D.; Lindahl, E.; Hess, B.; GROMACS Development Team. GROMACS User Manual Version 2018. Available online: www.gromacs.org (accessed on 2 November 2020).
54. Jorgensen, W.L.; Tirado-Rives, J. The OPLS [optimized potentials for liquid simulations] potential functions for proteins, energy minimizations for crystals of cyclic peptides and crambin. *J. Am. Chem. Soc.* **1988**, *110*, 1657–1666. [CrossRef] [PubMed]
55. Jorgensen, W.L.; Maxwell, D.S.; Tirado-Rives, J. Development and testing of the OPLS all-atom force field on conformational energetics and properties of organic liquids. *J. Am. Chem. Soc.* **1996**, *118*, 11225–11236. [CrossRef]
56. Jorgensen, W.L.; McDonald, N.A. Development of an all-atom force field for heterocycles. Properties of liquid pyridine and diazenes. *J. Mol. Struct. Theochem* **1998**, *424*, 145–155. [CrossRef]
57. McDonald, N.A.; Jorgensen, W.L. Development of an All-Atom Force Field for Heterocycles. Properties of Liquid Pyrrole, Furan, Diazoles, and Oxazoles. *J. Phys. Chem. B* **1998**, *102*, 8049–8059. [CrossRef]
58. Rizzo, R.C.; Jorgensen, W.L. OPLS All-Atom Model for Amines: Resolution of the Amine Hydration Problem. *J. Am. Chem. Soc.* **1999**, *121*, 4827–4836. [CrossRef]
59. Price, M.L.P.; Ostrovsky, D.; Jorgensen, W.L. Gas-phase and liquid-state properties of esters, nitriles, and nitro compounds with the OPLS-AA force field. *J. Comput. Chem.* **2001**, *22*, 1340–1352. [CrossRef]
60. Watkins, E.K.; Jorgensen, W.L. Perfluoroalkanes: Conformational Analysis and Liquid-State Properties from ab Initio and Monte Carlo Calculations. *J. Phys. Chem. A* **2001**, *105*, 4118–4125. [CrossRef]
61. Kaminski, G.A.; Friesner, R.A.; Tirado-Rives, J.; Jorgensen, W.L. Evaluation and Reparametrization of the OPLS-AA Force Field for Proteins via Comparison with Accurate Quantum Chemical Calculations on Peptides. *J. Phys. Chem. B* **2001**, *105*, 6474–6487. [CrossRef]
62. Horn, H.W.; Swope, W.C.; Pitera, J.W.; Madura, J.D.; Dick, T.J.; Hura, G.L.; Head-Gordon, T. Development of an improved four-site water model for biomolecular simulations: TIP4P-Ew. *J. Chem. Phys.* **2004**, *120*, 9665–9678. [CrossRef]
63. Hohenberg, P.; Kohn, W. Inhomogeneous electron gas. *Phys. Rev.* **1964**, *136*, B864–B871. [CrossRef]
64. Kohn, W.; Sham, L.J. Self-Consistent Equations Including Exchange and Correlation Effects. *Phys. Rev.* **2002**, *140*, A1133–A1138. [CrossRef]
65. EMSL Basis Set Exchange. Available online: https://bse.pnl.gov/bse/portal (accessed on 5 November 2020).
66. Becke, A.D. Density-functional exchange-energy approximation with correct asymptotic behavior. *Phys. Rev. A* **1988**, *38*, 3098–3100. [CrossRef] [PubMed]
67. Lee, C.; Yang, W.; Parr, R.G. Development of the Colle-Salvetti correlation-energy formula into a functional of the electron density. *Phys. Rev. B* **1988**, *37*, 785–789. [CrossRef]
68. Valiev, M.; Bylaska, E.J.; Govind, N.; Kowalski, K.; Straatsma, T.P.; van Dam, H.J.J.; Wang, D.; Nieplocha, J.; Apra, E.; Windus, T.L.; et al. NWChem: A comprehensive and scalable open-source solution for large scale molecular simulations. *Comp. Phys. Commun.* **2010**, *181*, 1477–1489. [CrossRef]

69. Darden, T.A.; York, D.M.; Pedersen, L. Particle mesh Ewald: An N·log(N) method for Ewald sums in large systems. *J. Chem. Phys.* **1993**, *98*, 10089–10092. [CrossRef]
70. Essmann, U.; Perera, L.; Berkowitz, M.L.; Darden, T.; Lee, H.; Pedersen, L.G. A smooth particle mesh Ewald method. *J. Chem. Phys.* **1995**, *103*, 8577–8593. [CrossRef]
71. Páll, S.; Hess, B. A flexible algorithm for calculating pair interactions on SIMD architectures. *Comput. Phys. Commun.* **2013**, *184*, 2641–2650. [CrossRef]
72. Liebschner, D.; Afonine, P.V.; Baker, M.L.; Bunkóczi, G.; Chen, V.B.; Croll, T.I.; Hintze, B.; Hung, L.-W.; Jain, S.; McCoy, A.J.; et al. Macromolecular structure determination using X-rays, neutrons and electrons: Recent developments in Phenix. *Acta Crystallogr. Sect. D Struct. Biol.* **2019**, *75*, 861–877. [CrossRef]
73. Krissinel, E. Enhanced fold recognition using efficient short fragment clustering. *J. Mol. Biochem.* **2012**, *1*, 76–85.
74. Pang, Y.-P. At least 10% shorter C–H bonds in cryogenic protein crystal structures than in current AMBER forcefields. *Biochem. Biophys. Res. Commun.* **2015**, *458*, 352–355. [CrossRef] [PubMed]
75. Murshudov, G.N.; Skubák, P.; Lebedev, A.A.; Pannu, N.S.; Steiner, R.A.; Nicholls, R.A.; Winn, M.D.; Long, F.; Vagin, A.A. REFMAC5 for the refinement of macromolecular crystal structures. *Acta Crystallogr. Sect. D Biol. Crystallogr.* **2011**, *67*, 355–367. [CrossRef]
76. Engh, R.A.; Huber, R. Accurate bond and angle parameters for X-ray protein structure refinement. *Acta Crystallogr. Sect. A Found. Crystallogr.* **1991**, *47*, 392–400. [CrossRef]
77. Lovell, S.C.; Word, J.M.; Richardson, J.S.; Richardson, D.C. The penultimate rotamer library. *Proteins* **2000**, *40*, 389–408. [CrossRef]
78. Emsley, P.; Lohkamp, B.; Scott, W.G.; Cowtan, K. Features and development of Coot. *Acta Crystallogr. Sect. D Biol. Crystallogr.* **2010**, *66*, 486–501. [CrossRef]
79. Zhou, A.Q.; O'Hern, C.S.; Regan, L.J. Predicting the side-chain dihedral angle distributions of nonpolar, aromatic, and polar amino acids using hard sphere models. *Proteins Struct. Funct. Bioinform.* **2014**, *82*, 2574–2584. [CrossRef]
80. Hooft, R.W.W.; Vriend, G.; Sander, C.; Abola, E.E. Errors in protein structures. *Nature* **1996**, *381*, 272. [CrossRef]
81. Steinbach, P.J.; Brooks, B.R. New spherical-cutoff methods for long-range forces in macromolecular simulation. *J. Comput. Chem.* **1994**, *15*, 667–683. [CrossRef]
82. Da Silveira, C.H.; Pires, D.E.H.; Minardi, R.C.; Ribeiro, C.; Veloso, C.J.M.; Lopes, J.C.D.; Meira, W.; Neshich, G.; Ramos, C.H.I.; Habesch, R.; et al. Protein cutoff scanning: A comparative analysis of cutoff dependent and cutoff free methods for prospecting contacts in proteins. *Proteins* **2009**, *74*, 727–743. [CrossRef]
83. Reif, M.M.; Krautler, V.; Kastenholz, M.A.; Daura, X.; Hunenberger, P.H. Molecular dynamics simulations of a reversibly folding beta-heptapeptide in methanol: Influence of the treatment of long-range electrostatic interactions. *J. Phys. Chem. B* **2009**, *113*, 3112–3128. [CrossRef]
84. Piana, S.; Lindorff-Larsen, K.; Dirks, R.M.; Salmon, J.K.; Dror, R.O.; Shaw, D.E. Evaluating the Effects of Cutoffs and Treatment of Long-range Electrostatics in Protein Folding Simulations. *PLoS ONE* **2012**, *7*. [CrossRef] [PubMed]
85. Strzalka, R.; Buganski, I.; Smietanska, J.; Wolny, J. Structural Disorder in Quasicrystals. *Arch. Metall. Mater.* **2020**, *65*, 291–294.
86. Wolny, J.A.; Buganski, I.; Kuczera, P.; Strzałka, R. Pushing the limits of crystallography. *J. Appl. Crystallogr.* **2016**, *49*, 2106–2115. [CrossRef]
87. Buganski, I.; Strzalka, R.; Wolny, J. Phason-flips refinement of and multiple-scattering correction for the d-AlCuRh quasicrystal. *Acta Cryst.* **2019**, *A75*, 352–361. [CrossRef] [PubMed]
88. Mantsyzov, A.B.; Savelyev, O.Y.; Ivantcova, P.M.; Bräse, S.; Kudryavtsev, K.V.; Polshakov, V.I. Theoretical and NMR Conformational Studies of β-Proline Oligopeptides with Alternating Chirality of Pyrrolidine Units. *Front. Chem.* **2018**, *6*. [CrossRef] [PubMed]
89. Caballero, D.; Smith, W.W.; O'Hern, C.S.; Regan, L.J. Equilibrium transitions between side-chain conformations in leucine and isoleucine. *Proteins Struct. Funct. Bioinform.* **2015**, *83*, 1488–1499. [CrossRef] [PubMed]
90. Shapovalov, M.V.; Dunbrack, R.L. A Smoothed Backbone-Dependent Rotamer Library for Proteins Derived from Adaptive Kernel Density Estimates and Regressions. *Structure* **2011**, *19*, 844–858. [CrossRef]
91. Virrueta, A.; O'Hern, C.S.; Regan, L.J. Understanding the physical basis for the side-chain conformational preferences of methionine. *Proteins Struct. Funct. Bioinform.* **2016**, *84*, 900–911. [CrossRef]

MDPI
St. Alban-Anlage 66
4052 Basel
Switzerland
Tel. +41 61 683 77 34
Fax +41 61 302 89 18
www.mdpi.com

Crystals Editorial Office
E-mail: crystals@mdpi.com
www.mdpi.com/journal/crystals

www.ingramcontent.com/pod-product-compliance
Lightning Source LLC
LaVergne TN
LVHW070652170726
843515LV00004B/394

* 9 7 8 3 0 3 6 5 6 0 7 2 4 *